W9-ABN-171

Introduction to Singular Perturbations

APPLIED MATHEMATICS AND MECHANICS

An International Series of Monographs

EDITORS

Volume 1. K. Oswatitsch: Gas Dynamics, *English version by G. Kuerti* (1956)

Volume 2. G. Birkhoff and E. H. Zarantonello: Jet, Wakes, and Cavities (1957)

Volume 3. R. von Mises: Mathematical Theory of Compressible Fluid Flow, *Revised and completed by Hilda Geiringer and G. S. S. Ludford* (1958)

Volume 4. F. L. Alt: Electronic Digital Computers—Their Use in Science and Engineering (1958)

Volume 5A. Wallace D. Hayes and Ronald F. Probstein: Hypersonic Flow Theory, second edition, Volume I, Inviscid Flows (1966)

Volume 6. L. M. Brekhovskikh: Waves in Layered Media, *Translated from the Russian by D. Lieberman* (1960)

Volume 7. S. Fred Singer (ed.): Torques and Attitude Sensing in Earth Satellites (1964)

Volume 8. Milton Van Dyke: Perturbation Methods in Fluid Mechanics (1964)

Volume 9. Angelo Miele (ed.): Theory of Optimum Aerodynamic Shapes (1965)

Volume 10. Robert Betchov and William O. Criminale, Jr.: Stability of Parallel Flows (1967)

Volume 11. J. M. Burgers: Flow Equations for Composite Gases (1969)

Volume 12. John L. Lumley: Stochastic Tools in Turbulence (1970)

Volume 13. Henri Cabannes: Theoretical Magnetofluiddynamics (1970)

Volume 14. Robert E. O'Malley, Jr.: Introduction to Singular Perturbations (1974)

INTRODUCTION TO SINGULAR PERTURBATIONS

Robert E. O'Malley, Jr.

DEPARTMENT OF MATHEMATICS
UNIVERSITY OF ARIZONA
TUCSON, ARIZONA

ACADEMIC PRESS New York and London 1974
A Subsidiary of Harcourt Brace Jovanovich, Publishers

ACADEMIC PRESS, INC.
111 Fifth Avenue, New York, New York 10003

United Kingdom Edition published by
ACADEMIC PRESS, INC. (LONDON) LTD.
24/28 Oval Road, London NW1

Library of Congress Cataloging in Publication Data

O'Malley, Robert E
Introduction to singular perturbations.

(Applied mathematics and mechanics)
Bibliography: p.
1. Boundary value problems–Numerical solutions.
2. Perturbation Mathematics. 3. Asymptotic expansions. I. Title.
QA379.04 515'.35 73-9441
ISBN 0-12-525950-6
AMS(MOS) 1970 Subject Classifications: 34E15, 34E05, 34E20, 34A10, 34B05, 34B15, 35B30, 41A60, 49A10.

PRINTED IN THE UNITED STATES OF AMERICA

CONTENTS

PREFACE

Through this book the reader will become acquainted with certain fundamental techniques for obtaining asymptotic solutions to boundary value problems. It is intended for engineers and applied mathematicians, students of these disciplines, and others interested in the topic. It does not suppose many mathematical prerequisites.

Some knowledge of ordinary differential equations, linear algebra, and basic analysis is expected, however, as well as a belief in the inherent usefulness of the subject in important applications.

Topics covered are restricted partly by the experiences of the author and partly because it is his intention not to be too technical. Notably absent is any discussion of partial differential equations and of problems in unbounded domains. Without doubt, the underlying philosophy and methodology expressed here carries over with some modification to these and many other related areas of asymptotic analysis.

Further, it should be directly stated that the presentation is admittedly distorted by the current opinions and biases of the author. Instead of using inner and outer approximations, for example, outer expansions are corrected in regions of nonuniform convergence by the addition of "boundary layer corrections." This works most satisfactorily for the problems discussed. No claim of universal applicability is intended, nor is there any intention to denigrate the common practice of using local expansions to study complicated physical problems.

One of the most exciting features of singular perturbations is its unexpected appearance in many varied areas of application, often disguised among certain intuitive practices which are part of the folkways of the specific field. Quite naturally, then, singular perturbation techniques are among the basic tools of many applied scientists, but the common content of their experiences is not well known.

Among the literature of singular perturbations are the books of Van Dyke, Wasow, Kaplun, Cole, Nayfeh, and Eckhaus. Each has a different outlook, as does this volume. Nearly two hundred references have been cited, but this represents only a small portion of the relevant literature. Several hundred more items are to be found in the literature.

Earlier versions of this work have been used several times as the basis of a one-semester course for applied mathematics and engineering students at New York University. Students were encouraged to work through the detailed calculations indicated and to obtain solutions to examples illustrating the results obtained here or in references cited. A second course was then offered as a seminar with students giving presentations on singular perturbation methods for partial differential equations and on applications of interest to them.

Thanks are due to many individuals who have taught me about singular perturbations through their publications, lectures, conversations, or questions. Many more should be thanked for their interest in this and the encouagement they provided. The writing of this manuscript was supported by the Science Research Council at the University of Edinburgh, by the National Science Foundation (contract number GP-32996X2) and the Air Force Office of Scientific Research (contract number AFOSR 71-2013) at New York University, and by the Office of Naval Research (contract number N00014-67-A-0209-0022) at the University of Arizona.

It is hoped that these notes will be useful and interesting to a wide variety of scientists.

CHAPTER 1

FIRST CONCERNS

1. EXAMPLES OF SINGULAR PERTURBATION PROBLEMS

Consider a family of boundary value problems P_ε depending on a small parameter ε. Under many conditions, a "solution" $y_\varepsilon(x)$ of P_ε can be constructed by the well-known "method of perturbation"—i.e., as a power series in ε with first term y_0 being the solution of the problem P_0. When such an expansion converges as $\varepsilon \to 0$ uniformly in x, we have a regular perturbation problem. When $y_\varepsilon(x)$ does not have a uniform limit in x as $\varepsilon \to 0$, this regular perturbation method will fail and we have a singular perturbation problem. We shall list examples of such problems.

In these examples and below, it will be convenient to use the Landau order symbols O and o which are defined as follows: Given two functions $f(\varepsilon)$ and $g(\varepsilon)$, we write $f = O(g)$ as $\varepsilon \to 0$ if $|f(\varepsilon)/g(\varepsilon)|$ is bounded as $\varepsilon \to 0$. We write $f = o(g)$ as $\varepsilon \to 0$ if $f(\varepsilon)/g(\varepsilon) \to 0$ as $\varepsilon \to 0$. [For further discussion, see Olver (1973) or Erdélyi (1956).]

EXAMPLE 1: Let P_ε be the initial value problem

$$\begin{aligned} y'' + (1+\varepsilon)y &= 0, \qquad 0 \le x < \infty \\ y(0) &= 0, \qquad y'(0) = 1. \end{aligned} \tag{1.1}$$

Here, the "reduced problem,"

$$\begin{aligned} y'' + y &= 0 \\ y(0) &= 0, \qquad y'(0) = 1, \end{aligned}$$

obtained by setting $\varepsilon = 0$, has the solution $y_0(x) = \sin x$. A power series expansion

$$\tilde{y}_\varepsilon(x) = \sum_{j=0}^{\infty} y_j(x)\varepsilon^j$$

can be easily constructed to formally solve (1.1) termwise. Equating coefficients of ε^j successively in the differential equation and the initial conditions, each y_j for $j \ge 1$ will satisfy

$$\begin{aligned} y''_j + y_j &= -y_{j-1} \\ y_j(0) &= y'_j(0) = 0. \end{aligned}$$

Any number of terms y_j can obviously be constructed recursively. [For example, $y_1(x) = \frac{1}{2}(x \cos x - \sin x)$.] The actual solution $y_\varepsilon(x)$, however, is given by

$$y_\varepsilon(x) = (1+\varepsilon)^{-1/2} \sin\left((1+\varepsilon)^{1/2} x\right).$$

In order to compare y_ε with the partial sums of $\tilde{y}_\varepsilon$ formally obtained, we expand

$$(1+\varepsilon)^{1/2} = \sum_{k=0}^{\infty} C_k \varepsilon^k = 1 + \tfrac{1}{2}\varepsilon - \tfrac{1}{8}\varepsilon^2 + \cdots$$

for small ε and note that the mean value theorem implies that

$$y_\varepsilon(x) = \left(\sum_{k=0}^{N} C_k \varepsilon^k\right)^{-1} \sin\left(\left(\sum_{k=0}^{N} C_k \varepsilon^k\right)x\right) + O(\varepsilon^{N+1})$$

whenever $\varepsilon^{N+1}x = o(1)$ as $\varepsilon \to 0$. Thus, ordinary Taylor series expansion of the right-hand side yields

$$y_\varepsilon(x) = \sum_{j=0}^{N} y_j(x)\varepsilon^j + O(\varepsilon^{N+1}), \tag{1.2}$$

whenever x is restricted such that $\varepsilon^{N+1}x = o(1)$ as $\varepsilon \to 0$ (and, in particular, on bounded x intervals for ε sufficiently small). (The reader should do the expansion to check that the coefficients y_j are those found previously by using the differential equation.) Applying the mean value theorem, however, shows that (1.2) is no longer valid when x is so large compared to ε^{-1} that $1/x = o(\varepsilon^{N+1})$. Thus, we have a regular perturbation problem if we restrict x to bounded intervals. If we let x become appropriately unbounded as $\varepsilon \to 0$, however, the initial value problem can no longer be solved by the regular perturbation method since ultimately (1.2) no longer holds. The reader might have expected this breakdown in uniformity since $y_1(x)$, for example, does not remain bounded as $x \to \infty$. [Those familiar with two-variable expansions (cf. Section 1.4) will realize that an expansion using "times" t and εt can be formally generated, but this too will fail to remain valid for all $t \geq 0$.] Thus we observe that there is not necessarily any universal validity to formally obtained expansions.

EXAMPLE 2: The two-point problem

$$\begin{aligned} y'' - \varepsilon^2 y = 0, &\qquad 0 \leq x < \infty, \quad \varepsilon > 0 \\ y(0) = 1, &\qquad y(\infty) = 0, \end{aligned} \tag{1.3}$$

has the solution $y_\varepsilon(x) = e^{-\varepsilon x}$. Here

$$\lim_{\varepsilon \to 0} y_\varepsilon(x) = \begin{cases} 1 & \text{if} \quad \varepsilon x = o(1) \\ e^{-c} & \text{if} \quad \varepsilon x = c + o(1), \quad c \text{ arbitrary} \\ 0 & \text{if} \quad \varepsilon x \to \infty. \end{cases}$$

For x bounded, of course, $y_\varepsilon \to 1$ as $\varepsilon \to 0$, but there is no uniform limit for $x \geq 0$. Note that setting $\varepsilon = 0$ in (1.3) yields a reduced problem

$$y'' = 0$$
$$y(0) = 1, \qquad y(\infty) = 0,$$

with no solution $y_0(x)$. Further, if we introduce $\sigma = \varepsilon x$, the resulting problem

$$y_{\sigma\sigma} - y = 0, \qquad y(0) = 1, \qquad y(\infty) = 0$$

has the unique solution

$$y(\sigma) = e^{-\sigma} \qquad \text{for all} \quad \sigma \geq 0.$$

As might be anticipated, introduction of appropriate new coordinates is a basic technique used in obtaining solutions to singular perturbation problems.

The nonuniform convergence exhibited in these two examples occurred because the independent variable ranged over the infinite interval $x \geq 0$. Moreover, the nonuniformities "occurred at $x = \infty$." In many singular perturbation problems x is restricted to bounded domains. In the following, this shall generally be so. Instead, the problems are singular because the order of the differential equation defining P_ε for ε positive drops when ε becomes zero. Nonanalytic dependence of the solution y_ε on ε (as in the next example) often results in markedly different behavior, depending on how $\varepsilon \to 0$. Henceforth, then, *we shall take ε to be a small positive parameter* or a small complex parameter defined in a narrow sector about the positive real ε axis.

Before proceeding, we note that

$$\frac{\varepsilon}{x + \varepsilon} \qquad \text{and} \qquad e^{-x/\varepsilon}$$

are examples of functions which converge nonuniformly at $x = 0$ as $\varepsilon \to 0$. For $x > 0$, both functions tend to zero as $\varepsilon \to 0$, while at

$x = 0$ both are equal to one. (See Figs. 1 and 2.) Functions featuring analogous nonuniform convergence are typical elements of solutions to singular perturbations problems.

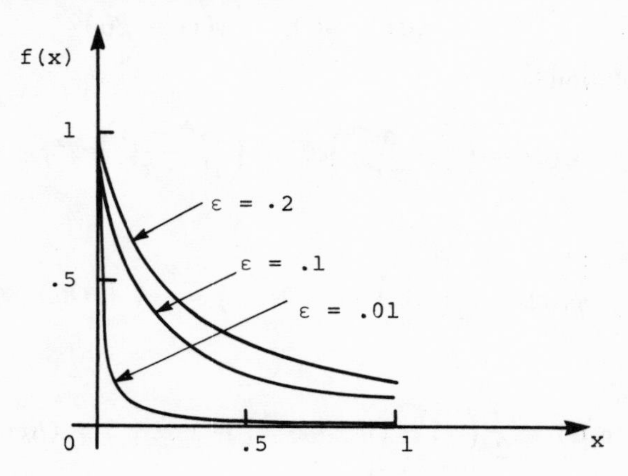

FIGURE 1 The function $f(x) = \varepsilon/(x + \varepsilon)$ for $\varepsilon = 0.2, 0.1, 0.01$.

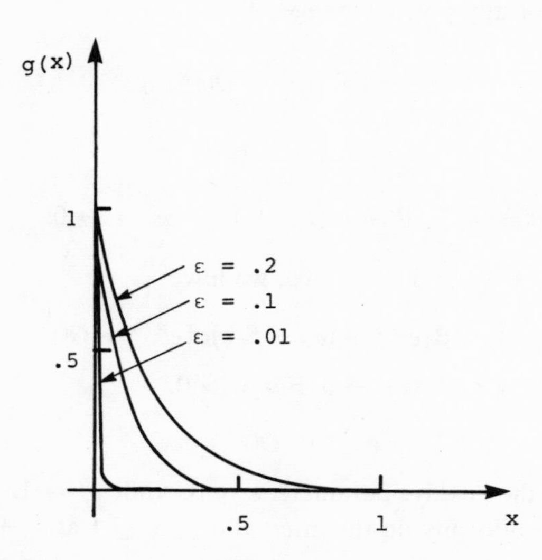

FIGURE 2 The function $g(x) = e^{-x/\varepsilon}$ for $\varepsilon = 0.2, 0.1, 0.01$.

EXAMPLE 3: The two-point problem

$$\begin{aligned} \varepsilon y'' + y' + y &= 0, \qquad 0 \le x \le 1 \\ y(0) &= \alpha(\varepsilon), \qquad y(1) = \beta(\varepsilon) \end{aligned} \tag{1.4}$$

has the solution

$$y_\varepsilon(x) = \left(\frac{\beta - \alpha e^{\rho_2}}{e^{\rho_1} - e^{\rho_2}}\right)e^{\rho_1 x} + \left(\frac{\alpha e^{\rho_1} - \beta}{e^{\rho_1} - e^{\rho_2}}\right)e^{\rho_2 x},$$

where

$$\rho_1(\varepsilon) = \frac{1}{2\varepsilon}\left(-1 + (1 - 4\varepsilon)^{1/2}\right) = -1 + O(\varepsilon)$$

and

$$\rho_2(\varepsilon) = \frac{1}{2\varepsilon}\left(-1 - (1 - 4\varepsilon)^{1/2}\right) = -\frac{1}{\varepsilon} + 1 + O(\varepsilon)$$

are roots of the characteristic equation

$$\varepsilon\rho^2 + \rho + 1 = 0.$$

Suppose that for any positive integer N

$$\alpha(\varepsilon) = \sum_{j=0}^{N} \alpha_j \varepsilon^j + O(\varepsilon^{N+1})$$

and

$$\beta(\varepsilon) = \sum_{j=0}^{N} \beta_j \varepsilon^j + O(\varepsilon^{N+1}) \qquad \text{as} \quad \varepsilon \to 0.$$

Then, since $\rho_1 \to -1$ and $\rho_2 \to -\infty$, we have

$$y_\varepsilon(x) = \beta_0 e^{1-x} + (\alpha_0 - \beta_0 e)e^x e^{-x/\varepsilon} + O(\varepsilon) \tag{1.5}$$

throughout $0 \le x \le 1$ as $\varepsilon \to 0$. For $x > 0$,

$$e^{-x/\varepsilon} = O(\varepsilon^N)$$

for every N as the positive parameter $\varepsilon \to 0$, while $e^0 = 1$. Thus, $y_\varepsilon(x)$ converges nonuniformly on the interval $0 \le x \le 1$ as $\varepsilon \to 0$ and

$$y_\varepsilon(x) = \beta_0 e^{1-x} + O(\varepsilon) \qquad \text{for} \quad x > 0.$$

Note that the "limiting solution"

$$Y_0(x) = \beta_0 e^{1-x}$$

satisfies the "reduced problem"

$$\begin{aligned} y' + y &= 0 \\ y(1) &= \beta_0 \end{aligned} \tag{1.6}$$

and that

$$y_\varepsilon(x) \to Y_0(x) \qquad \text{as} \quad \varepsilon \to 0$$

except at $x = 0$, where $y_\varepsilon(0) = \alpha$. This nonuniform convergence as $\varepsilon \to 0$ (unless $\alpha_0 = \beta_0 e$) implies that we have a singular perturbation problem. (See Fig. 3.) In general, then,

$$\begin{aligned} \lim_{x\to 0}\left(\lim_{\varepsilon\to 0} y_\varepsilon(x)\right) &= Y_0(0) \\ &\neq \lim_{\varepsilon\to 0}\left(\lim_{x\to 0} y_\varepsilon(x)\right) = \alpha_0 . \end{aligned}$$

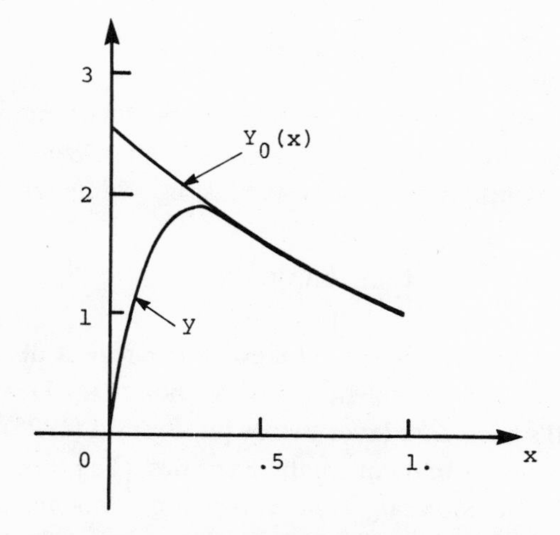

FIGURE 3 Nonuniform convergence of the solution y of $\varepsilon y'' + y' + y = 0$, $y(0) = 0$, $y(1) = 1$ to the solution $Y_0(x) = e^{1-x}$ of the reduced problem.

Higher-order approximations may also be obtained. Specifically, we have

$$y_\varepsilon(x) = [\beta(\varepsilon)e^{-(1-x)\rho_1(\varepsilon)}] + \left[\left(\alpha(\varepsilon) - \beta(\varepsilon)e^{-\rho_1(\varepsilon)}\right)\left(e^{(\rho_2(\varepsilon)+1/\varepsilon)x}\right)\right]e^{-x/\varepsilon} + O(\varepsilon^N) \qquad \text{for every integer} \quad N \geq 0.$$

Expanding the bracketed expressions as power series in ε, we obtain approximations of the form

$$y_\varepsilon(x) = \left(\sum_{j=0}^{N} A_j(x)\varepsilon^j\right) + \left(\sum_{j=0}^{N} B_j(x)\varepsilon^j\right)e^{-x/\varepsilon} + O(\varepsilon^{N+1}) \tag{1.7}$$

uniformly on the closed interval [0, 1] while

$$y_\varepsilon(x) = \sum_{j=0}^{N} A_j(x)\varepsilon^j + O(\varepsilon^{N+1})$$

uniformly on any interval $0 < \delta \leq x \leq 1$ as $\varepsilon \to 0$.

The limiting solution away from $x = 0$,

$$\sum_{j=0}^{\infty} A_j(x)\varepsilon^j, \tag{1.8}$$

must be a formal power series solution of the equation $\varepsilon y'' + y' + y = 0$. (Thus, $A_0' + A_0 = 0$, $A_1' + A_1 = -A_0''$, etc.) We shall call this sum the "outer solution" or "outer expansion," while we shall call

$$\left(\sum_{j=0}^{\infty} B_j(x)\varepsilon^j\right)e^{-x/\varepsilon} \tag{1.9}$$

the "boundary layer correction." Likewise the region of nonuniform convergence near $x = 0$ will be called a "boundary layer" in reference to Prandtl's boundary layer theory [cf. Prandtl (1905) and Meyer (1971)] for flow past a body at small viscosities. [The well-read should be warned that the boundary layer correction is not the same as the inner (or boundary layer) solution used by many authors, including Kaplun (1967) and Van Dyke (1964).]

Why the limiting solution $Y_0(x) = A_0(x) = \beta_0 e^{1-x}$ retains the boundary condition at $x = 1$ (or, equivalently, why the nonuniform convergence occurs at $x = 0$) needs to be explained. We note, for now, that the limiting solution would satisfy the limiting boundary condition at $x = 0$ if, instead, ε were small and negative.

EXAMPLE 4: We now consider the nonlinear problem

$$\begin{aligned} \varepsilon y'' &= yy', \qquad -1 \leq x \leq 1 \\ y(-1) &= \alpha, \qquad y(1) = \beta, \end{aligned} \tag{1.10}$$

as the real boundary values α and β vary remaining independent of ε. Before proceeding, note that any solution of this two-point problem will be unique by the maximum principle [cf. Protter and Weinberger (1967) and Dorr *et al.* (1973)]. Six cases will be considered separately.

CASE 1: $\alpha > 0$, $\alpha + \beta > 0$. Here the reader can verify that the solution is given by

$$y_\varepsilon(x) = \alpha \left[\frac{1 + \left(1 - \dfrac{2\alpha}{\alpha + \beta}\right) e^{(\alpha/\varepsilon)(x-1)}}{1 - \left(1 - \dfrac{2\alpha}{\alpha + \beta}\right) e^{(\alpha/\varepsilon)(x-1)}} \right] \tag{1.11}$$

up to asymptotically exponentially small terms, so

$$y_\varepsilon(x) \to \alpha \qquad \text{as} \quad \varepsilon \to 0 \quad \text{for} \quad x < 1. \tag{1.12}$$

Since $y(1) = \beta$, nonuniform convergence generally occurs at $x = 1$.

CASE 2: $\beta < 0$, $\alpha + \beta < 0$. Knowing the solution in Case 1, we readily find that

$$y_\varepsilon(x) = \beta \left[\frac{1 + \left(1 - \dfrac{2\beta}{\alpha + \beta}\right) e^{(\beta/\varepsilon)(x+1)}}{1 - \left(1 - \dfrac{2\beta}{\alpha + \beta}\right) e^{(\beta/\varepsilon)(x+1)}} \right] \tag{1.13}$$

up to asymptotically exponentially small terms and

$$y_\varepsilon(x) \to \beta \quad \text{as} \quad \varepsilon \to 0 \quad \text{for} \quad x > -1 \tag{1.14}$$

with nonuniform convergence at $x = -1$.

CASE 3: $\alpha = -\beta > 0$. Here

$$y_\varepsilon(x) = -\alpha\left[\frac{1 - e^{-x\alpha/\varepsilon}}{1 + e^{-x\alpha/\varepsilon}}\right] \tag{1.15}$$

up to asymptotically exponentially small terms. Thus

$$y_\varepsilon(x) \to \begin{cases} \alpha & \text{for} \quad -1 \leq x < 0 \\ 0 & \text{for} \quad x = 0 \\ \beta & \text{for} \quad 0 < x \leq 1 \end{cases} \tag{1.16}$$

as $\varepsilon \to 0$. The nonuniformity at the interior point $x = 0$ might be considered roughly reminiscent of shock phenomena in fluid mechanics [cf., e.g., Cole (1951) and Murray (1973)].

CASE 4: $\alpha < 0 < \beta$. The function

$$y_\varepsilon(x) = \varepsilon\pi\left(1 + \frac{\varepsilon(\beta - \alpha)C(\varepsilon)}{\alpha\beta}\right)$$

$$\times \tan\left[\frac{\pi}{2}\left(\left(1 + \frac{\varepsilon(\beta - \alpha)C(\varepsilon)}{\alpha\beta}\right)x - \frac{\varepsilon(\beta + \alpha)k(\varepsilon)}{\alpha\beta}\right)\right] \tag{1.17}$$

solves the problem for unique constants $C(\varepsilon) = 1 + O(\varepsilon)$ and $k(\varepsilon) = 1 + O(\varepsilon)$. Thus

$$y_\varepsilon(x) \to 0 \quad \text{as} \quad \varepsilon \to 0 \quad \text{for} \quad -1 < x < 1 \tag{1.18}$$

and nonuniform convergence occurs near both $x = \pm 1$.

CASE 5: $\alpha = 0, \beta > 0$. Here

$$y_\varepsilon(x) = \frac{\varepsilon\pi}{2}\left(1 - \frac{\varepsilon\, m(\varepsilon)}{\beta}\right) \tan\left[\frac{\pi}{4}\left(1 - \frac{\varepsilon\, m(\varepsilon)}{\beta}\right)(x+1)\right] \tag{1.19}$$

for the appropriate $m(\varepsilon) = 1 + O(\varepsilon)$. Thus

$$y_\varepsilon(x) \to 0 \quad \text{as} \quad \varepsilon \to 0 \quad \text{for} \quad -1 \leq x < 1 \tag{1.20}$$

and nonuniform convergence occurs at $x = 1$.

CASE 6: $\beta = 0, \alpha < 0$. Similar to Case 5, we have

$$y_\varepsilon(x) = \frac{\varepsilon\pi}{2}\left(1 + \frac{\varepsilon\, n(\varepsilon)}{\alpha}\right) \tan\left[\frac{\pi}{4}\left(1 + \frac{\varepsilon\, n(\varepsilon)}{\alpha}\right)(x-1)\right] \tag{1.21}$$

for some $n(\varepsilon) = 1 + O(\varepsilon)$ and

$$y_\varepsilon(x) \to 0 \quad \text{as} \quad \varepsilon \to 0 \quad \text{for} \quad -1 < x \leq 1. \tag{1.22}$$

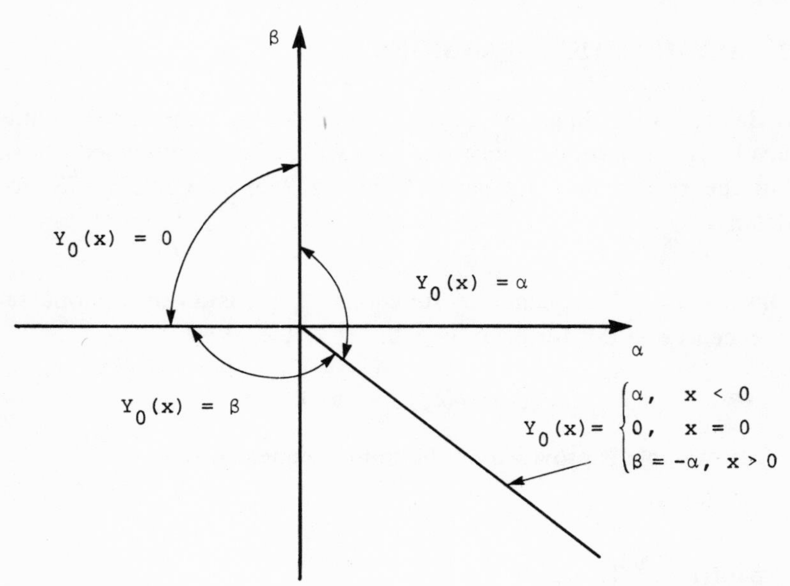

FIGURE 4 The dependence of the limiting solution $Y_0(x)$, $-1 < x < 1$, of $\varepsilon y'' = yy'$, $y(-1) = \alpha$, $y(1) = \beta$ on α and β.

Note that in all cases the limiting solution of (1.10), away from the nonuniformities, is constant and therefore satisfies the reduced equation $yy' = 0$. This limiting behavior within $-1 < x < 1$ is summarized in Fig. 4.

Similar analyses of the limiting behavior of the solution to

$$\varepsilon y'' + yy' - y = 0, \qquad 0 \leq x \leq 1$$

$$y(0) = \alpha, \qquad y(1) = \beta$$

have been done by several authors [see, especially, Dorr *et al.* (1973)]. The problem is more difficult than (1.10), but the asymptotic results are similar. Further examples which illustrate "the capriciousness of singular perturbations" are contained in the article by Wasow (1970), which is recommended reading. More complicated examples, treated heuristically, are discussed by Carrier (1974).

2. ASYMPTOTIC EXPANSIONS

Before proceeding, the reader should know some basic results involving asymptotic expansions. They will be briefly discussed below, but the reader should consult Erdélyi (1956) or Olver (1973) for details.

DEFINITION: A sequence of functions $\{\phi_n(\varepsilon)\}$ is an asymptotic sequence as $\varepsilon \to 0$ if for every $n \geq 0$

$$\phi_{n+1} = o(\phi_n) \qquad \text{as} \quad \varepsilon \to 0.$$

An example is provided by the power sequence where

$$\phi_n(\varepsilon) = \varepsilon^n.$$

DEFINITION: The series

$$\sum_n A_n \phi_n(\varepsilon) \tag{1.23}$$

is an asymptotic approximation (to N terms) of the function $f(\varepsilon)$ as $\varepsilon \to 0$ with respect to the asymptotic sequence $\{\phi_n\}$ if

$$f(\varepsilon) = \sum_{n=0}^{N} A_n \phi_n(\varepsilon) + o\big(\phi_N(\varepsilon)\big) \qquad \text{as} \quad \varepsilon \to 0. \tag{1.24}$$

If (1.24) holds for every $N \geq 0$, we write

$$f(\varepsilon) \sim \sum_{n=0}^{\infty} A_n \phi_n(\varepsilon) \tag{1.25}$$

and call (1.25) the asymptotic expansion of $f(\varepsilon)$ as $\varepsilon \to 0$.

The reader should realize that convergent series expansions are asymptotic, while asymptotic series expansions are, in general, divergent. Only the finite sums of (1.24) need be defined for small enough ε; no infinite series need be summed.

DEFINITION: The functions $f(\varepsilon)$ and $g(\varepsilon)$ are asymptotically equal with respect to the asymptotic sequence $\{\phi_n(\varepsilon)\}$ if

$$f(\varepsilon) = g(\varepsilon) + o\big(\phi_n(\varepsilon)\big) \text{ as } \quad \varepsilon \to 0$$

for all integers $n \geq 0$. In particular, we write

$$f(\varepsilon) \sim 0$$

if $f(\varepsilon) = o(\phi_n(\varepsilon))$ as $\varepsilon \to 0$ for all n, and we then call $f(\varepsilon)$ asymptotically negligible.

As an example, note that $e^{-1/\varepsilon} \sim 0$ with respect to the power sequence $\{\varepsilon^n\}$ as $\varepsilon \to 0$ through positive values.

Note further that the coefficients of the asymptotic expansion (1.23) are uniquely determined since, for each m,

$$A_m = \lim_{\varepsilon \to 0} \left(\frac{f(\varepsilon) - \sum_{n=0}^{m-1} A_n \phi_n(\varepsilon)}{\phi_m(\varepsilon)} \right). \tag{1.26}$$

Linear combinations and integrals of asymptotic expansions are defined in the obvious manner. Differentiation of asymptotic expansions is not always possible since, for example, the function

$$f(\varepsilon) = e^{-1/\varepsilon} \sin e^{1/\varepsilon}$$

satisfies $f(\varepsilon) \sim 0$ for the asymptotic sequence $\{\phi_n(\varepsilon)\} = \{\varepsilon^n\}$ as ε tends to zero through positive real values. However,

$$f'(\varepsilon) = \frac{1}{\varepsilon^2}[e^{-1/\varepsilon} \sin e^{1/\varepsilon} - \cos e^{1/\varepsilon}]$$

does not have a limit as $\varepsilon \to 0$ and does not possess an asymptotic expansion with respect to $\{\varepsilon^n\}$. Likewise, multiplication of asymptotic expansions is generally not possible because elements of the double sequence $\{\phi_m \phi_n\}$ cannot always be expanded with respect to the single asymptotic sequence $\{\phi_p\}$.

We shall usually restrict attention to asymptotic power series

$$\sum_{r=0}^{\infty} A_r \varepsilon^r. \tag{1.27}$$

Then multiplication is always possible and termwise differentiation of the expansion of any function $f(\varepsilon)$ is permitted provided $f(\varepsilon)$ is holomorphic in any complex sector

$$S = \{\varepsilon\colon 0 < |\varepsilon| \le \varepsilon_0,\ \theta_1 \le \arg \varepsilon \le \theta_2,\ \theta_1 < \theta_2\}$$

[cf. Wasow (1965, p. 39)]. Further, for asymptotic power series, we have the *Borel-Ritt theorem* [cf. Wasow (1965, p. 43)]:

Given any formal power series (1.27) *and any sector* S, *there is a (nonunique) function* $f(\varepsilon)$ *holomorphic in* S *for* $|\varepsilon|$ *arbitrarily small so that*

$$f(\varepsilon) \sim \sum_{r=0}^{\infty} A_r \varepsilon^r \qquad \text{as} \quad \varepsilon \to 0 \quad \text{in} \quad S.$$

This basic result will be important in following chapters for summing asymptotic series which are formally obtained. For generalizations of this theorem, see Pittnauer (1969, 1972).

Often it is necessary to use asymptotic expansions more general than the Poincaré expansions introduced above. Alternative definitions have been given by Schmidt (1937), van der Corput (1955–1956), Erdélyi (1961b), Erdélyi and Wyman (1963), and Riekstins (1966).

DEFINITION (ERDÉLYI AND WYMAN): The formal series

$$\sum_{k=0}^{\infty} f_k(\varepsilon) \tag{1.28}$$

is an asymptotic expansion of the function $f(\varepsilon)$ with respect to the asymptotic sequence $\{\psi_j\}$ as $\varepsilon \to 0$ if for every $N \geq 0$

$$f(\varepsilon) = \sum_{k=0}^{N} f_k(\varepsilon) + o\big(\psi_N(\varepsilon)\big) \qquad \text{as} \quad \varepsilon \to 0. \tag{1.29}$$

We write

$$f(\varepsilon) \sim \sum_{k=0}^{\infty} f_k(\varepsilon) \qquad \{\psi_j\} \quad \text{as} \quad \varepsilon \to 0. \tag{1.30}$$

Note that expansions of this general form are not unique. The nonuniqueness allows greater flexibility and extends considerably the class of functions whose asymptotic behavior can be studied. The unique Poincaré expansions correspond to the special case where $f_k(\varepsilon) = A_k \psi_k(\varepsilon)$.

3. INTUITIVE APPROACH OF MATCHED ASYMPTOTIC EXPANSIONS

The method of matched asymptotic expansions (or of inner and outer expansions) has been successfully applied to numerous physical problems which involve singular perturbations. The basic ideas involved were present in Prandtl's boundary layer theory and in other even earlier work [cf. the historical survey of Van Dyke (1971)]; they occurred in Friedrichs' work in elasticity [see Friedrichs and Stoker (1941) and Friedrichs (1950)] and in his mathematical lectures (Friedrichs, 1953, 1955); but they became best known through the work of Kaplun and Lagerstrom [cf. Kaplun (1967) and Lagerstrom and

Casten (1972)] and through the book of Van Dyke (1964). Despite some uncertainties [cf. Fraenkel (1969)], the variant technique which Van Dyke called the asymptotic matching principle has been most widely accepted by users. Meanwhile, work continues toward the development of a theory based on the concept of overlap domains and Kaplun's extension theorem [see, e.g., Meyer (1967) and Eckhaus (1973)]. Among the most significant mathematical singular perturbations theory, Vasil'eva's work [cf. the survey articles by Vasil'eva (1962, 1963a,b), and Wasow (1965)] closely parallels the popular matching techniques. Matching methods have generally superseded earlier techniques of patching solutions (often numerically) at a somewhat arbitrary point, presumably in an "overlap domain."

In any perturbation problem involving a small positive parameter ε, it is natural to seek a solution of the form

$$y_\varepsilon^{\mathrm{o}}(x) \sim \sum_{j=0}^{\infty} a_j(x)\beta_j(\varepsilon) \qquad \text{as} \quad \varepsilon \to 0, \tag{1.31}$$

where x ranges over some (usually bounded) domain D and $\{\beta_j(\varepsilon)\}$ is an asymptotic sequence (often the power sequence $\{\varepsilon^j\}$) as $\varepsilon \to 0$. In a singular perturbation problem, such an expansion cannot be valid uniformly in x. (For example, this solution may fail to satisfy all boundary conditions. In applications, moreover, physical considerations will often indicate which boundary conditions are so omitted.) Instead, the expansion $y_\varepsilon^{\mathrm{o}}$ will be generally satisfactory in the "outer region" away from (part of) the boundary of D. It will be called an outer expansion (or outer solution). In order to investigate regions of nonuniform convergence, one introduces one or more stretching transformations

$$\xi = \psi(x, \varepsilon) \tag{1.32}$$

which "blow up" a region of nonuniformity (near a part of the boundary with neglected boundary conditions, for example). Thus $\xi = x/\varepsilon$ might be used for nonuniform convergence at $x = 0$. Then if ξ is fixed and $\varepsilon \to 0$, $x \to 0$, while if $x > 0$ is fixed and $\varepsilon \to 0$, $\xi \to \infty$.

Selection of "correct" stretching transformations is an art sometimes motivated by physical considerations (cf. Van Dyke's discussion on reference lengths), or mathematically (as in Kaplun's concept of principal limits). In terms of the stretched variable ξ, one might seek an asymptotic solution of the form

$$y_\varepsilon^{\mathrm{i}}(\xi) \sim \sum_{j=0}^{\infty} b_j(\xi)\alpha_j(\varepsilon) \qquad \text{as} \quad \varepsilon \to 0 \tag{1.33}$$

(where the sequence $\{\alpha_j(\varepsilon)\}$ is asymptotic as $\varepsilon \to 0$) valid for values of ξ in some "inner region." This will be called an inner expansion. (The inner expansion often accounts for boundary conditions neglected by the outer expansion.) [We note that this generally accepted "inner–outer" terminology is natural to boundary layer flow problems in fluid dynamics. In the theory of thin elastic shells, however, exactly the opposite terminology is appropriate and used; cf. Gol'denveizer (1961) and Reiss (1962).] The inner region will generally shrink completely as $\varepsilon \to 0$ when expressed in terms of the outer variable x. Hence, the inner expansion is "local."

In most problems, it is impossible to determine both the outer and the inner expansions $y_\varepsilon^{\mathrm{o}}$ and $y_\varepsilon^{\mathrm{i}}$ completely by straightforward expansion procedures. Since both expansions should represent the solution of the original problem asymptotically in different regions, one might attempt to "match" them, i.e., to formally relate the outer expansion in the inner region $(y_\varepsilon^{\mathrm{o}})^{\mathrm{i}}$ and the inner expansion in the outer region $(y_\varepsilon^{\mathrm{i}})^{\mathrm{o}}$ through use of the stretching $\xi = \psi(x, \varepsilon)$. The rules for even formally accomplishing this, in all generality, can be very complicated (cf. Fraenkel). Justification in particular examples through use of an overlap domain and intermediate limits is difficult (cf. Lagerstrom and Casten) and, in general, there is no *a priori* reason to believe that an overlap domain (where matching is possible) exists. Once matching is accomplished, however, the asymptotic solution to well-posed problems becomes completely known in both the inner and outer regions.

Frequently, it is convenient to obtain a composite expansion y^{c} uniformly valid in D. One method of doing so is to let

$$y_\varepsilon^{\mathrm{c}} = y_\varepsilon^{\mathrm{o}} + y_\varepsilon^{\mathrm{i}} - (y_\varepsilon^{\mathrm{i}})^{\mathrm{o}},$$

making the appropriate modification if several regions of nonuniform convergence (several inner regions) are necessary. [Note that $(y_\varepsilon^{\rm i})^{\rm i} = y_\varepsilon^{\rm i}$, so that $(y_\varepsilon^{\rm c})^{\rm i} = y_\varepsilon^{\rm i}$. Likewise, in the outer region, $(y_\varepsilon^{\rm c})^{\rm o} = y_\varepsilon^{\rm o}$.] In certain problems, as Van Dyke observes, other composite expansions ("multiplicative," for example) may be preferable.

We refer the reader to the cited references for more details of this important technique and many examples of its application. Most of the techniques used in following chapters to construct asymptotic solutions to singular perturbation problems can be interpreted as specialized matched expansion methods, although we shall not generally so identify them.

EXAMPLE: Let us reconsider the simple problem

$$\varepsilon y'' + y' + y = 0$$

$$y(0) = \alpha_0, \qquad y(1) = \beta_0$$

of Section 1.1 and obtain an asymptotic approximation to the solution by the method of matched asymptotic expansions. We shall seek an outer expansion as the asymptotic power series

$$y_\varepsilon^{\rm o}(x) \sim \sum_{j=0}^{\infty} a_j(x)\varepsilon^j$$

valid for $0 < x \leq 1$, and an inner expansion

$$y_\varepsilon^{\rm i}(\xi) \sim \sum_{j=0}^{\infty} b_j(\xi)\varepsilon^j$$

in terms of the stretched variable

$$\xi = x/\varepsilon$$

valid as $\varepsilon \to 0$ near $x = 0$.

The outer expansion must satisfy the differential equation and the terminal condition asymptotically; i.e.,

$$\varepsilon(a_0'' + \varepsilon a_1'' + \cdots) + (a_0' + \varepsilon a_1' + \varepsilon^2 a_2' + \cdots)$$
$$+ (a_0 + \varepsilon a_1 + \varepsilon^2 a_2 + \cdots) = 0$$

and

$$a_0(1) + \varepsilon a_1(1) + \varepsilon^2 a_2(1) + \cdots = \beta_0.$$

Thus, equating coefficients,

$$a_0' + a_0 = 0, \qquad a_0(1) = \beta_0$$
$$a_1' + a_1 + a_0'' = 0, \qquad a_1(1) = 0,$$

etc., so that

$$a_0(x) = \beta_0 e^{1-x}$$

and

$$a_1(x) = (1 - x)\beta_0 e^{1-x}.$$

In terms of ξ, note that the differential equation becomes

$$y_{\xi\xi} + y_\xi + \varepsilon y = 0$$

since

$$\frac{d}{dx} = \frac{1}{\varepsilon}\frac{d}{d\xi} \qquad \text{and} \qquad \frac{d^2}{dx^2} = \frac{1}{\varepsilon^2}\frac{d^2}{d\xi^2}.$$

Thus, the inner expansion $y_\varepsilon^{\mathrm{i}}(\xi)$ must asymptotically satisfy this differential equation and the initial condition; i.e.,

$$(b_{0\xi\xi} + \varepsilon b_{1\xi\xi} + \varepsilon^2 b_{2\xi\xi} + \cdots) + (b_{0\xi} + \varepsilon b_{1\xi} + \varepsilon^2 b_{2\xi} + \cdots)$$
$$+ \varepsilon(b_0 + \varepsilon b_1 + \cdots) = 0$$

and

$$b_0(0) + \varepsilon b_1(0) + \varepsilon^2 b_2(0) + \cdots = \alpha_0.$$

Then

$$b_{0\xi\xi} + b_{0\xi} = 0, \qquad b_0(0) = \alpha_0$$
$$b_{1\xi\xi} + b_{1\xi} + b_0 = 0, \qquad b_1(0) = 0,$$

etc., so integration yields

$$b_0(\xi) = \gamma_0 + (\alpha_0 - \gamma_0)e^{-\xi}$$

and

$$b_1(\xi) = -(\gamma_0\xi + \gamma_1) + [(\alpha_0 - \gamma_0)\xi + \gamma_1]e^{-\xi}$$

with γ_0 and γ_1 being undetermined constants. Hence,

$$y_\varepsilon^{\text{o}}(x) = \beta_0 e^{1-x} + \varepsilon(1-x)\beta_0 e^{1-x} + O(\varepsilon^2)$$

while

$$\begin{aligned} y_\varepsilon^{\text{i}}(x) = [\gamma_0 + (\alpha_0 - \gamma_0)e^{-\xi}] + \varepsilon[-(\gamma_0\xi + \gamma_1) \\ + \Big((\alpha_0 - \gamma_0)\xi + \gamma_1\Big)e^{-\xi}] + O(\varepsilon^2). \end{aligned}$$

Writing the outer expansion in terms of the inner variable ξ, then, we have

$$y_\varepsilon^{\text{o}}(x) = \beta_0 e e^{-\varepsilon\xi} + \varepsilon(1 - \varepsilon\xi)\beta_0 e e^{-\varepsilon\xi} + O(\varepsilon^2)$$

which leads to the $O(\varepsilon^2)$ approximation

$$\left(y_\varepsilon^{\text{o}}(x)\right)^{\text{i}} \approx \beta_0 e\Big(1 + \varepsilon(-\xi + 1)\Big).$$

Analogously, writing the inner expansion in terms of the outer variable x for $x > 0$, the $e^{-\xi}$ terms are asymptotically negligible and we have the $O(\varepsilon^2)$ approximation

$$\left(y_\varepsilon^{\text{i}}(x)\right)^{\text{o}} \approx \gamma_0(1 - x) - \varepsilon\gamma_1.$$

Since $\xi = x/\varepsilon$, matching (to this order) will be accomplished by selecting

$$\gamma_0 = -\gamma_1 = \beta_0 e.$$

Expressing all approximations in terms of the outer variable x, then, we have the composite approximation

$$
\begin{aligned}
y_\varepsilon^{\mathrm{c}}(x) \approx{}& [\beta_0 e^{1-x} + \varepsilon(1-x)\beta_0 e^{1-x}] + \Big[\beta_0 e + (\alpha_0 - \beta_0 e)e^{-x/\varepsilon} \\
&+ \varepsilon\Big(-\beta_0 e\Big(\frac{x}{\varepsilon} - 1\Big) + \Big((\alpha_0 - \beta_0 e)\frac{x}{\varepsilon} - \beta_0 e\Big)e^{-x/\varepsilon}\Big)\Big] \\
&- \Big[\beta_0 e + \varepsilon\beta_0 e\Big(-\frac{x}{\varepsilon} + 1\Big)\Big]
\end{aligned}
$$

or

$$
\begin{aligned}
y_\varepsilon^{\mathrm{c}}(x) \approx{}& [\beta_0 e^{1-x} + (\alpha_0 - \beta_0 e)(1+x)e^{-x/\varepsilon}] \\
&+ \varepsilon[(1-x)\beta_0 e^{1-x} - \beta_0 e e^{-x/\varepsilon}],
\end{aligned}
$$

which should be compared with the exact solution previously given. Higher-order approximations could also be obtained analogously. The inner and outer expansions are depicted in Fig. 5. The asymptotic solution (cf. Fig. 3) follows the inner solution near $x = 0$ and the outer solution for $x > 0$.

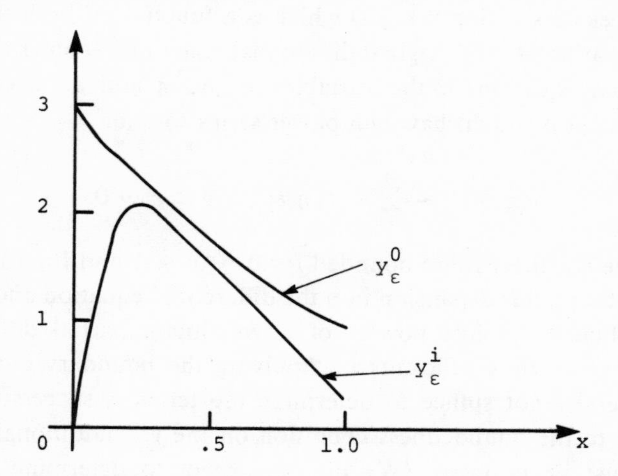

FIGURE 5 The inner expansion $y_\varepsilon^{\mathrm{i}}$ and the outer expansion $y_\varepsilon^{\mathrm{o}}$ for $\varepsilon y'' + y' + y = 0$, $y(0) = 0$, $y(1) = 1$, $\varepsilon = 0.1$.

4. TWO-VARIABLE EXPANSIONS

Two-variable expansion techniques have frequently been used to solve initial value problems with a small parameter on semi-infinite intervals [cf. Mahony (1961–1962), Kevorkian (1966), Cole (1968), Kollett (1973), and Nayfeh (1973)]. For such problems, "two-timing" is closely related to well-known averaging methods [cf. Morrison (1966), Perko (1969), and Whitham (1970)].

Analogous techniques have also been applied to obtain asymptotic solutions to singularly perturbed boundary value problems for ordinary differential equations on finite intervals, say $0 \leq x \leq 1$ [cf. Cochran (1962), Erdélyi (1968b), O'Malley (1968a, 1970a), and Searl (1971)]. In addition to the ("slow-time") variable x, one introduces another ("fast-time") variable η ranging over an unbounded interval. [If, for example, nonuniform convergence occurred at $x = x_0$, the fast variable $\eta = (1/\varepsilon) \int_{x_0}^{x} g(s)\, ds$ for some positive g might be appropriate.] Selection of a proper fast variable may be based on different physical time scales occurring, may be motivated by simple model equations, or may be left somewhat arbitrary initially.

One seeks a solution $y(x, \eta, \varepsilon)$ which is a function of both the slow and fast variables. The original differential equation becomes a partial differential equation in the variables x and η and an asymptotic solution of it is sought having a power series solution

$$y(x, \eta, \varepsilon) \sim \sum_{j=0}^{\infty} y_j(x, \eta)\varepsilon^j \qquad \text{as} \quad \varepsilon \to 0, \tag{1.34}$$

where the coefficients are bounded for $0 \leq x \leq 1$ and for all $\eta \geq 0$.

Substituting this expansion into the differential equation and equating coefficients of like powers of ε, we obtain partial differential equations for the coefficients y_j. Applying the boundary conditions will generally not suffice to determine the terms y_j successively. In addition to the boundedness condition on the y_j's, additional conditions must be imposed. (We do not expect to determine the y_j's uniquely, since the expansion sought is a generalized asymptotic expansion, but we need to eliminate some arbitrariness.) Following

Poincaré, one usually asks that certain "secular terms" (like, e.g., $\eta^k e^{-\eta}$, $k > 0$) be eliminated. These somewhat arbitrary requirements can be motivated mathematically. For example, Erdélyi and Searl seek approximations whose correctness as $\varepsilon \to 0$ can be ascertained by applying their previously obtained theorems. Analogously, Reiss (1971) seeks to minimize the error at each step in the approximation.

A special, but frequently occurring, situation is when the two-variable expansion has the form

$$y(x, \eta, \varepsilon) = y_1(x, \varepsilon) + y_2(\eta, \varepsilon),$$

i.e., when there is an additive decomposition of the asymptotic solution as the sum of functions of the slow and fast variables separately. Indeed, the ("additive") composite expansions resulting from matched asymptotic expansions generally have this form.

It is important to also observe that equations with slowly varying coefficients arise in a variety of physical applications and engineering approximations. Two-time techniques are well suited to such problems.

EXAMPLE: As an illustration of the two-variable method, consider the two-point problem

$$\varepsilon y'' + a(x)y' + b(x)y = 0$$

$$y(0) = \alpha_0, \qquad y(1) = \beta_0,$$

where a and b are infinitely differentiable functions on $0 \leq x \leq 1$ and $a(x) > 0$ there. By analogy to the constant coefficient problem treated previously, one might expect nonuniform convergence as $\varepsilon \to 0$ near $x = 0$ and that the fast variable

$$\eta = \frac{1}{\varepsilon} \int_0^x a(s)\, ds$$

might be appropriate. (Many other choices are also possible.) Proceeding with this η requires the solution $y(x, \eta, \varepsilon)$ to satisfy the partial differential equation

$$\frac{a^2(x)}{\varepsilon}[y_{\eta\eta} + y_\eta] + [2a(x)y_{x\eta} + a'(x)y_\eta + a(x)y_x + b(x)y] + \varepsilon y_{xx} = 0$$

since

$$\frac{d}{dx} = \frac{\partial}{\partial x} + \frac{a(x)}{\varepsilon}\frac{\partial}{\partial \eta}$$

and

$$\frac{d^2}{dx^2} = \frac{\partial^2}{\partial x^2} + \frac{2a(x)}{\varepsilon}\frac{\partial^2}{\partial x\,\partial \eta} + \frac{a'(x)}{\varepsilon}\frac{\partial}{\partial \eta} + \frac{a^2(x)}{\varepsilon^2}\frac{\partial^2}{\partial \eta^2}.$$

Substituting the expansion

$$y(x,\eta,\varepsilon) \sim \sum_{j=0}^{\infty} y_j(x,\eta)\varepsilon^j$$

into this differential equation, we formally equate coefficients of each power of ε separately to zero. Thus, from the coefficient of ε^{-1}, we have

$$a^2(x)[y_{0\eta\eta} + y_{0\eta}] = 0$$

and, integrating, we obtain

$$y_0(x,\eta) = A_0(x) + C_0(x)e^{-\eta},$$

where A_0 and C_0 are undetermined. Likewise, from the coefficient of ε^0, we have

$$a^2(x)[y_{1\eta\eta} + y_{1\eta}] + 2a(x)y_{0x\eta} + a'(x)y_{0\eta} + a(x)y_{0x} + b(x)y_0 = 0.$$

Integrating with respect to η, then,

$$a^2(x)[y_{1\eta} + y_1] + 2a(x)y_{0x} + a'(x)y_0 + \int^{\eta}[a(x)y_{0x} + b(x)y_0]\,d\eta = 0$$

and, substituting for y_0, we have

$$a^2(x)[y_{1\eta} + y_1] + \big(a(x)A_0' + b(x)A_0\big)\eta$$
$$+ \big(a(x)C_0' + a'(x)C_0 - b(x)C_0\big)e^{-\eta}$$
$$= a^2(x)A_1(x)$$

for A_1 arbitrary. We then integrate with respect to η. Since y_1 must remain bounded as $\eta \to \infty$ (i.e., as $\varepsilon \to 0$ for $x > 0$), we must have

$$a(x)A_0' + b(x)A_0 = 0.$$

Likewise, to avoid a secular term in y_1 which is a multiple of $\eta e^{-\eta}$, we must have

$$a(x)C_0' + a'(x)C_0 - b(x)C_0 = 0.$$

Thus, integration implies that y_1 has the form

$$y_1(x, \eta) = A_1(x) + C_1(x)e^{-\eta},$$

where A_1 and C_1 are so far arbitrary. Applying the boundary conditions to $y = A_0(x) + C_0(x)e^{-\eta} + \varepsilon(\cdots)$, noting that $e^{-\eta}$ is asymptotically negligible at $x = 0$, implies that we must select

$$A_0(1) = \beta_0 \qquad \text{and} \qquad C_0(0) = \alpha_0 - A_0(0).$$

Summarizing, then, we have begun to develop an asymptotic solution of the form

$$y(x, \eta, \varepsilon) = \big(A_0(x) + C_0(x)e^{-\eta}\big) + \varepsilon\big(A_1(x) + C_1(x)e^{-\eta}\big) + O(\varepsilon^2),$$

where

$$A_0(x) = \beta_0 \exp\left(-\int_1^x \frac{b(s)}{a(s)}\,ds\right)$$

$$C_0(x) = \frac{a(0)}{a(x)}\big(\alpha_0 - A_0(0)\big)\exp\left(\int_0^x \frac{b(s)}{a(s)}\,ds\right)$$

and A_1 and C_1 are, as yet, undetermined. We note that it was necessary to determine the form of the second-order coefficient y_1 (by boundedness and secularity conditions) in order to completely obtain the first coefficient y_0. This is typically the case for two-time methods.

This formal result can be continued to higher-order terms and can be easily shown to be asymptotically correct by comparison, say, with the asymptotic solution constructed in Section 3.1 by other methods.

CHAPTER 2

THE REGULAR PERTURBATION METHOD

To illustrate the regular perturbation method (which will be basic to the singular perturbation techniques developed later), we consider the nonlinear initial value problem:

$$\begin{aligned} \frac{dy}{dx} &= f(x, y, \varepsilon) \\ y(x_0) &= c(\varepsilon), \end{aligned} \tag{2.1}$$

where f and c have asymptotic power series expansions

$$f(x, y, \varepsilon) \sim \sum_{j=0}^{\infty} f_j(x, y)\varepsilon^j$$

and

$$c(\varepsilon) \sim \sum_{j=0}^{\infty} c_j \varepsilon^j$$

as the small parameter ε tends to zero and the f_j's are infinitely differentiable in x and y. (In this discussion, there is no need to restrict ε to any sector.) We shall assume that the "reduced problem"

$$\begin{aligned} \frac{dy}{dx} &= f_0(x,y) \\ y(x_0) &= c_0 \end{aligned} \tag{2.2}$$

has a unique solution $y_0(x)$ on the bounded interval $|x - x_0| \leq B$, $B > 0$, and we shall seek a solution $y(x, \varepsilon)$ of the "full problem" (2.1) having an asymptotic expansion of the form

$$y(x, \varepsilon) \sim \sum_{j=0}^{\infty} y_j(x)\varepsilon^j \qquad \text{as} \quad \varepsilon \to 0. \tag{2.3}$$

Formal termwise differentiation implies

$$f\big(x, y(x, \varepsilon), \varepsilon\big) = \frac{dy}{dx}(x, \varepsilon) \sim \sum_{j=0}^{\infty} \frac{dy_j}{dx}\varepsilon^j,$$

while substitution and termwise rearrangement yields

$$f\big(x, y(x, \varepsilon), \varepsilon\big) \sim \sum_{j=0}^{\infty} f^j(x, y_0, y_1, \ldots, y_j)\varepsilon^j,$$

where

$$\begin{aligned} f^0(x, y_0) &= f_0(x, y_0) \\ f^1(x, y_0, y_1) &= y_1 f_{0y}(x, y_0) + f_1(x, y_0) \end{aligned}$$

and, generally, for each $j \geq 1$,

$$f^j(x, y_0, y_1, \ldots, y_j) = y_j f_{0y}(x, y_0) + \tilde{f}_{j-1}(x),$$

where $\tilde{f}_{j-1}$ depends on x, $y_0(x), \ldots, y_{j-1}(x)$ only. Likewise,

$$c(\varepsilon) \sim \sum_{j=0}^{\infty} y_j(x_0)\varepsilon^j.$$

Equating coefficients of like powers of ε in the differential equation and the initial condition, then, we ask that y_0 satisfy the nonlinear initial value problem

$$\frac{dy_0}{dx} = f_0(x, y_0)$$

$$y_0(x_0) = c_0,$$

while successive y_j's satisfy the linear initial value problems

$$\begin{aligned}\frac{dy_j}{dx} &= f^j(x, y_0, \ldots, y_j)\\ &= y_j f_{0y}(x, y_0) + \tilde{f}_{j-1}(x)\\ y_j(x_0) &= c_j.\end{aligned}$$

Hence, $y_0(x)$ is the unique solution of the reduced problem (2.2) on $|x - x_0| \leq B$ and integration implies that the y_j's are successively given by

$$y_j(x) = c_j \exp\left[\int_{x_0}^{x} f_{0y}\big(s, y_0(s)\big)\, ds\right] + \int_{x_0}^{x} \exp\left[\int_{t}^{x} f_{0y}\big(s, y_0(s)\big)\, ds\right] \tilde{f}_{j-1}(t)\, dt.$$

Thus, a unique expansion (2.3) has been formally obtained.

To prove the asymptotic correctness, we set

$$y(x, \varepsilon) = \sum_{j=0}^{N} y_j(x)\varepsilon^j + R(x, \varepsilon), \tag{2.4}$$

where N is any nonnegative integer. We shall show that the remainder R is unique and $O(\varepsilon^{N+1})$ throughout $|x - x_0| \leq B$ as $\varepsilon \to 0$. This will prove that the problem (2.1) has a unique solution $y(x, \varepsilon)$ on $|x - x_0| \leq B$ for ε sufficiently small whose asymptotic series expansion is given by (2.3) there.

Differentiating (2.4), we first note that

$$\frac{dy}{dx} = \sum_{j=0}^{N} f^j(x, y_0, \ldots, y_j)\varepsilon^j + \frac{dR}{dx}.$$

However,

$$f\left(x, \sum_{j=0}^{N} y_j(x)\varepsilon^j + R(x,\varepsilon), \varepsilon\right) = \sum_{j=0}^{N} f^j(x, y_0, \ldots, y_j)\varepsilon^j + \tilde{f}(x, R, \varepsilon),$$

where $\tilde{f}$ is of the form

$$\tilde{f}(x, R, \varepsilon) = R\hat{f}_0(x, \varepsilon) + \varepsilon^{N+1}\hat{f}_1(x, \varepsilon) + R^2\hat{f}_2(x, R, \varepsilon),$$

where $\hat{f}_0$, $\hat{f}_1$, and $\hat{f}_2$ are smooth functions of their arguments. Since y satisfies (2.1), it follows that

$$\frac{dR}{dx} = \tilde{f}(x, R, \varepsilon) \qquad \text{for} \quad |x - x_0| \leq B$$

$$R(x_0, \varepsilon) = c(\varepsilon) - \sum_{j=0}^{N} c_j \varepsilon^j \sim \varepsilon^{N+1} \sum_{j=0}^{\infty} c_{N+1+j}\varepsilon^j.$$

Integrating, R must satisfy the nonlinear Volterra integral equation

$$R(x, \varepsilon) = R^0(x, \varepsilon) + \int_{x_0}^{x} R^2(t, \varepsilon)\hat{f}_2\big(t, R(t, \varepsilon), \varepsilon\big) \times \exp\left[\int_t^x \hat{f}_0(s, \varepsilon)\, ds\right] dt$$

on $|x - x_0| \leq B$, where

$$R^0(x, \varepsilon) = R(x_0, \varepsilon)\exp\left[\int_{x_0}^{x} \hat{f}_0(s, \varepsilon)\, ds\right] + \varepsilon^{N+1}\int_{x_0}^{x} \hat{f}_1(t, \varepsilon)\exp\left[\int_t^x \hat{f}_0(s, \varepsilon)\, ds\right] dt = O(\varepsilon^{N+1}).$$

This equation, however, can be uniquely solved throughout $|x - x_0|$

$\leq B$ for ε sufficiently small. We merely recursively define

$$R^j(x, \varepsilon) = R^0(x, \varepsilon) + \int_{x_0}^{x} \left(R^{j-1}(t, \varepsilon)\right)^2 \hat{f}_2\left(t, R^{j-1}(t, \varepsilon), \varepsilon\right) \times \exp\left[\int_t^x \hat{f}_0(s, \varepsilon)\, ds\right] dt = O(\varepsilon^{N+1})$$

for each $j \geq 1$ and the unique solution is given by

$$R(x, \varepsilon) = \lim_{j \to \infty} R^j(x, \varepsilon) = O(\varepsilon^{N+1}), \qquad |x - x_0| \leq B.$$

This follows by writing

$$R^j(x, \varepsilon) = \sum_{k=1}^{j} \left(R^k(x, \varepsilon) - R^{k-1}(x, \varepsilon)\right)$$

and by proving convergence to $R(x, \varepsilon)$ in the space of continuous functions on $|x - x_0| \leq B$ by estimating the differences $R^k(x, \varepsilon) - R^{k-1}(x, \varepsilon)$ [cf. Erdélyi (1964) and Willett (1964)]. Summarizing, then, we have

THEOREM 1: *Suppose the initial value problem*

$$\frac{dy}{dx} = f_0(x, y)$$
$$y(x_0) = c_0$$

has a unique solution $y_0(x)$ throughout $|x - x_0| \leq B$, B finite. Then, for each ε sufficiently small, the problem

$$\frac{dy}{dx} = f(x, y, \varepsilon) \sim \sum_{j=0}^{\infty} f_j(x, y)\varepsilon^j$$
$$y(x_0) = c(\varepsilon) \sim \sum_{j=0}^{\infty} c_j \varepsilon^j$$

with infinitely differentiable coefficients $f_j(x, y)$ has a unique bounded solution for $|x - x_0| \leq B$ such that for each $N \geq 0$

$$y(x, \varepsilon) = \sum_{j=0}^{N} y_j(x)\varepsilon^j + O(\varepsilon^{N+1})$$

there. Further, the coefficients $y_j, j \geq 1$, may be uniquely obtained recursively as solutions of linear initial value problems.

NOTE

1. Less differentiability of the coefficients f_j requires the expansion for $y(x, \varepsilon)$ to be terminated after an appropriate number of terms.

2. The theorem also holds when y is a vector.

3. The theorem is false on unbounded intervals (cf. Example 1, Section 1.1) without additional hypotheses.

The reader should be aware that regular perturbation techniques find applications in a wide variety of settings and levels of sophistication [cf., e.g., Bellman (1964), Rellich (1969), Morse and Feshbach (1953), and Kato (1966)].

EXAMPLE: Consider the concrete example

$$\frac{dy}{dx} = m(x, \varepsilon)y^2 + n(x, \varepsilon)y + p(x, \varepsilon)$$

$$y(0) = c(\varepsilon) \sim \sum_{j=0}^{\infty} c_j \varepsilon^j,$$

where coefficients m, n, and p have asymptotic series expansions as $\varepsilon \to 0$ with infinitely differentiable coefficients; e.g.,

$$m(x, \varepsilon) \sim \sum_{j=0}^{\infty} m_j(x)\varepsilon^j.$$

Suppose further that the reduced problem

$$\frac{dy}{dx} = m_0(x)y^2 + n_0(x)y + p_0(x)$$

$$y(0) = c_0$$

has a unique solution $y_0(x)$ on some interval $|x| \leq B$, B finite.

Seeking an asymptotic series expansion for the solution,

$$y(x, \varepsilon) \sim \sum_{j=0}^{\infty} y_j(x)\varepsilon^j,$$

and formally substituting into the differential equation, we have

$$\begin{aligned}
\frac{dy_0}{dx} + \varepsilon\frac{dy_1}{dx} + \varepsilon^2\frac{dy_2}{dx} + \cdots
&= (m_0 + \varepsilon m_1 + \varepsilon^2 m_2 + \cdots) \\
&\quad \times (y_0^2 + 2\varepsilon y_0 y_1 + \varepsilon^2[2y_0 y_2 + y_1^2] + \cdots) \\
&\quad + (n_0 + \varepsilon n_1 + \varepsilon^2 n_2 + \cdots)(y_0 + \varepsilon y_1 + \varepsilon^2 y_2 + \cdots) \\
&\quad + (p_0 + \varepsilon p_1 + \varepsilon^2 p_2 + \cdots) \\
&= (m_0 y_0^2 + n_0 y_0 + p_0) + \varepsilon(2m_0 y_0 y_1 + m_1 y_0^2 + n_0 y_1 + n_1 y_0 + p_1) \\
&\quad + \varepsilon^2(2m_0 y_0 y_2 + m_0 y_1^2 + 2m_1 y_0 y_1 + m_2 y_0^2 + n_0 y_2 + n_1 y_1 \\
&\quad + n_2 y_0 + p_2) + \cdots.
\end{aligned}$$

Thus, equating coefficients here and in the initial condition, we have

$$\begin{aligned}
\frac{dy_0}{dx} &= m_0 y_0^2 + n_0 y_0 + p_0, \qquad y_0(0) = c_0 \\
\frac{dy_1}{dx} &= (2m_0 y_0 + n_0)y_1 + (m_1 y_0^2 + n_1 y_0 + p_1), \qquad y_1(0) = c_1 \\
\frac{dy_2}{dx} &= (2m_0 y_0 + n_0)y_2 \\
&\quad + (m_0 y_1^2 + 2m_1 y_0 y_1 + m_2 y_0^2 + n_1 y_1 + n_2 y_0 + p_2), \\
&\qquad\qquad y_2(0) = c_2,
\end{aligned}$$

etc. Taking y_0 as the unique solution of the nonlinear reduced problem and integrating the linear equations for y_1 and y_2, we have

$$y_1(x) = c_1 \exp\left[\int_0^x \left(2m_0(s)y_0(s) + n_0(s)\right) ds\right] + \int_0^x \left(m_1(t)y_0^2(t) + n_1(t)y_0(t) + p_1(t)\right) \times \exp\left[\int_t^x \left(2m_0(s)y_0(s) + n_0(s)\right) ds\right] dt$$

and

$$y_2(x) = c_2 \exp\left[\int_0^x \left(2m_0(s)y_0(s) + n_0(s)\right) ds\right] + \int_0^x \left(m_0(t)y_1^2(t) + 2m_1(t)y_0(t)y_1(t) + m_2(t)y_0^2(t) + n_1(t)y_1(t) + n_2(t)y_0(t) + p_2(t)\right) \times \exp\left[\int_t^x \left(2m_0(s)y_0(s) + n_0(s)\right) ds\right] dt.$$

Further terms can likewise be easily obtained recursively on $|x| \leq B$ (where y_0 is defined). A specific example is pictured in Fig. 6.

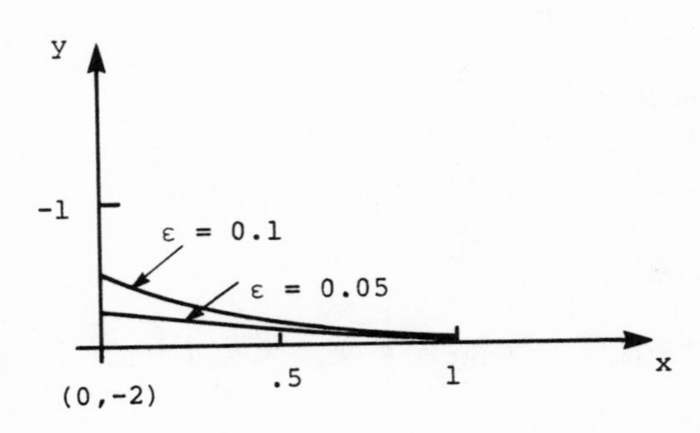

FIGURE 6 The solution $y(x, \varepsilon) = -2 + 5\varepsilon e^{-2x} + O(\varepsilon^2)$ of $dy/dx = y^2 + 2y$, $y(0) = -2 + 5\varepsilon$.

CHAPTER 3

LINEAR BOUNDARY VALUE PROBLEMS

1. SECOND-ORDER PROBLEMS

In this section we shall consider boundary value problems for the linear equation

$$\varepsilon y'' + a(x)y' + b(x)y = 0 \tag{3.1}$$

on the interval $0 \le x \le 1$ as the small positive parameter ε tends to zero. We suppose that $a(x)$ and $b(x)$ are infinitely differentiable and that

$$a(x) \ge \delta > 0 \tag{3.2}$$

there. [If $a(x) < 0$, the substitution $z = 1 - x$ will allow this condition to be satisfied for $0 \le z \le 1$.] Due to the linearity, all solutions will be linear combinations of any two linearly independent solutions [cf., e.g., Coddington and Levinson (1955)]. Analyzing constant coefficient examples (like Example 3 of Section 1.1) prompts us to seek linearly independent asymptotic solutions of the form

$$y_1(x, \varepsilon) = A(x, \varepsilon) \tag{3.3a}$$

and

$$y_2(x, \varepsilon) = B(x, \varepsilon) \exp\left[-\frac{1}{\varepsilon}\int_0^x a(s)\, ds\right], \tag{3.3b}$$

where

$$A(x, \varepsilon) \sim \sum_{j=0}^{\infty} a_j(x)\varepsilon^j, \qquad A(0, \varepsilon) = 1 \tag{3.4a}$$

and

$$B(x, \varepsilon) \sim \sum_{j=0}^{\infty} b_j(x)\varepsilon^j, \qquad B(0, \varepsilon) = 1. \tag{3.4b}$$

Note that $y_1(x, \varepsilon)$ will be bounded throughout $0 \le x \le 1$ while $y_2(x, \varepsilon)$ will be exponentially small as $\varepsilon \to 0$ for $x > 0$. Substituting (3.3) into the differential equation implies that we must have

$$a(x)A' + b(x)A = -\varepsilon A''$$

and

$$a(x)B' + a'(x)B - b(x)B = \varepsilon B''$$

throughout $0 \le x \le 1$. Then, using the expansions (3.4) and equating coefficients of like powers of ε, we find that the a_j's and b_j's are uniquely determined recursively as follows:

$$\begin{aligned} a_0(x) &= \exp\left[-\int_0^x \frac{b(s)}{a(s)}\, ds\right] \\ a_j(x) &= -\int_0^x \frac{a''_{j-1}(t)}{a(t)} \exp\left[-\int_t^x \frac{b(s)}{a(s)}\, ds\right] dt, \qquad j \ge 1 \end{aligned} \tag{3.5}$$

and

$$b_0(x) = \frac{a(0)}{a(x)} \exp\left[\int_0^x \frac{b(s)}{a(s)} ds\right]$$

$$b_j(x) = \frac{1}{a(x)} \int_0^x b''_{j-1}(t) \exp\left[\int_t^x \frac{b(s)}{a(s)} ds\right] dt, \qquad j \geq 1. \tag{3.6}$$

Recall the Borel–Ritt theorem of Section 1.2 which lets us define $\zeta(\varepsilon)$ and $\eta(\varepsilon)$ as analytic functions of ε such that

$$\zeta(\varepsilon) \sim \sum_{j=0}^{\infty} a'_j(0)\varepsilon^j$$

and

$$\eta(\varepsilon) \sim -a(0) + \varepsilon \sum_{j=0}^{\infty} b'_j(0)\varepsilon^j.$$

Thus, we have constructed a formal solution $A(x, \varepsilon)$ of the initial value problem

$$\begin{aligned} \varepsilon y'' + a(x)y' + b(x)y &= 0, \qquad 0 \leq x \leq 1 \\ y(0, \varepsilon) &= 1, \qquad y'(0, \varepsilon) = \zeta(\varepsilon) \end{aligned} \tag{3.7}$$

and a formal solution

$$B(x, \varepsilon) \exp\left[-\frac{1}{\varepsilon}\int_0^x a(s)\, ds\right]$$

of the initial value problem

$$\begin{aligned} \varepsilon y'' + a(x)y' + b(x)y &= 0, \qquad 0 \leq x \leq 1 \\ y(0, \varepsilon) &= 1, \qquad y'(0, \varepsilon) = \frac{\eta(\varepsilon)}{\varepsilon}. \end{aligned} \tag{3.8}$$

To proceed, we prove the asymptotic validity of these formal solutions:

LEMMA: *The initial value problem (3.7) has a unique solution $y_1(x,\varepsilon)$ such that, for each integer $N \geq 0$,*

$$y_1(x,\varepsilon) = \sum_{j=0}^{N} a_j(x)\varepsilon^j + \varepsilon^{N+1}R(x,\varepsilon), \tag{3.9}$$

where $R(x,\varepsilon)$ is bounded throughout $0 \leq x \leq 1$ for each $\varepsilon > 0$ sufficiently small. Likewise, the initial value problem (3.8) has a unique solution such that, for each integer $N \geq 0$,

$$y_2(x,\varepsilon) = \left(\sum_{j=0}^{N} b_j(x)\varepsilon^j\right) \exp\left[-\frac{1}{\varepsilon}\int_0^x a(s)\,ds\right] + \varepsilon^{N+1}S(x,\varepsilon), \tag{3.10}$$

where $S(x,\varepsilon)$ is bounded throughout $0 \leq x \leq 1$ for each $\varepsilon > 0$ sufficiently small.

PROOF: We shall prove the assertions about y_2 only, the results for y_1 being obtained in analogous fashion. Substituting (3.10) into (3.8), we find that $\tilde{S} = \varepsilon^{N+1}S$ must satisfy the initial value problem

$$\begin{aligned}
\varepsilon\tilde{S}'' + a(x)\tilde{S}' + b(x)\tilde{S} &= -\varepsilon^{N+1}b_N''(x), \qquad 0 \leq x \leq 1 \\
\tilde{S}(0,\varepsilon) &= 0 \\
\tilde{S}'(0,\varepsilon) &= \varepsilon^N\theta(\varepsilon) \\
&\equiv \frac{1}{\varepsilon}\left(\eta(\varepsilon) + a(0) - \sum_{j=0}^{N} b_j'(0)\varepsilon^j\right) \\
&\sim \varepsilon^N \sum_{j=0}^{\infty} b_{N+1+j}'(0)\varepsilon^j.
\end{aligned}$$

Integrating twice and reversing the order of integration, we find that $\tilde{S}(x,\varepsilon)$ will satisfy the linear Volterra integral equation

$$\tilde{S}(x,\varepsilon) = S^0(x,\varepsilon) - \int_0^x \left(\frac{1}{\varepsilon}\int_t^x \exp\left[-\frac{1}{\varepsilon}\int_t^r a(s)\,ds\right] dr\right) b(t)\,\tilde{S}(t,\varepsilon)\,dt,$$

where

$$S^0(x,\varepsilon) = \varepsilon^N \theta(\varepsilon) \int_0^x \exp\left[-\frac{1}{\varepsilon}\int_0^r a(s)\,ds\right] dr$$
$$- \varepsilon^{N+1} \int_0^x \left(\frac{1}{\varepsilon}\int_t^x \exp\left[-\frac{1}{\varepsilon}\int_t^r a(s)\,ds\right] dr\right) b_N''(t)\,dt.$$

Note that, since $a(x) \geq \delta > 0$,

$$\frac{1}{\varepsilon}\int_t^x \exp\left[-\frac{1}{\varepsilon}\int_t^r a(s)\,ds\right] dr \leq \frac{1}{\delta}\left[1 - \exp\left(-\frac{\delta}{\varepsilon}(x-t)\right)\right]$$

for $x \geq t$. This implies that $S^0(x,\varepsilon)$ is $O(\varepsilon^{N+1})$ throughout $[0, 1]$ and that the integral equation can be uniquely solved by successive approximations [cf. Erdélyi (1964)]. One defines

$$S^j(x,\varepsilon) = S^0(x,\varepsilon) - \int_0^x \left(\frac{1}{\varepsilon}\int_t^x \exp\left[-\frac{1}{\varepsilon}\int_t^r a(s)\,ds\right] dr\right)$$
$$\times\, b(t)\, S^{j-1}(t,\varepsilon)\,dt$$

for each $j \geq 1$, and the solution is given by

$$\tilde{S}(x,\varepsilon) = \lim_{j\to\infty} S^j(x,\varepsilon) = O(\varepsilon^{N+1}).$$

Alternatively, using the resolvent kernel $W(x,t,\varepsilon)$ corresponding to the Volterra kernel

$$K(x,t,\varepsilon) = -\frac{b(t)}{\varepsilon}\int_t^x \exp\left[-\frac{1}{\varepsilon}\int_t^r a(s)\,ds\right] dr,$$

note that we can represent the solution $\tilde{S}$ in the form

$$\tilde{S}(x,\varepsilon) = S^0(x,\varepsilon) + \int_0^x W(x,t,\varepsilon)\, S^0(t,\varepsilon)\,dt = O(\varepsilon^{N+1})$$

[cf. Cochran (1968)]. Since $\tilde{S}(x,\varepsilon) = O(\varepsilon^{N+1})$ is uniquely determined, the solution $y_2(x,\varepsilon)$ of (3.8) is also unique. This completes the proof.

Clearly, the two solutions y_1 and y_2 are linearly independent on $[0,1]$ for ε sufficiently small. Since we have constructed a fundamental set of asymptotic solutions of (3.1), the general solution will be of the form

$$y(x,\varepsilon) = k_1(\varepsilon)A(x,\varepsilon) + k_2(\varepsilon)B(x,\varepsilon)\exp\left[-\frac{1}{\varepsilon}\int_0^x a(s)\,ds\right] \tag{3.11}$$

for arbitrary functions $k_i(\varepsilon)$. Moreover, if any solution of (3.1) converges to a bounded limit in $0 \leq x \leq 1$ as $\varepsilon \to 0$, it must converge to the solution $k_1(0)\,a_0(x)$ of the reduced equation

$$a(x)y'(x) + b(x)y(x) = 0$$

for $x > 0$, and it will feature nonuniform convergence at $x = 0$ unless $k_2(0) = 0$. Derivatives of the solution will then also converge to the corresponding derivatives of $k_1(0)\,a_0(x)$ for $x > 0$.

Suppose, as a first example, that we wish to find the asymptotic solution of the boundary value problem

$$\begin{aligned} \varepsilon y'' + a(x)y' + b(x)y = 0, \qquad 0 \leq x \leq 1 \\ y(0) = \alpha(\varepsilon), \qquad y(1) = \beta(\varepsilon), \end{aligned} \tag{3.12}$$

where α and β have power series expansions as $\varepsilon \to 0$. The solution must be of the form (3.11) where $k_1(\varepsilon)$ and $k_2(\varepsilon)$ satisfy the linear equations

$$\begin{aligned} \alpha(\varepsilon) &= k_1(\varepsilon) + k_2(\varepsilon) \\ \beta(\varepsilon) &= k_1(\varepsilon)A(1,\varepsilon) + k_2(\varepsilon)B(1,\varepsilon)\exp\left[-\frac{1}{\varepsilon}\int_0^1 a(s)\,ds\right]. \end{aligned}$$

Since $A(1,\varepsilon) \neq 0$ and $a(x) \geq \delta > 0$, the system is nonsingular for ε sufficiently small and we have

$$k_1(\epsilon) = \frac{\beta(\varepsilon)}{A(1,\varepsilon)} + O(e^{-\delta/\varepsilon}),$$

and

$$k_2(\epsilon) = \alpha(\varepsilon) - \frac{\beta(\varepsilon)}{A(1,\epsilon)} + O(e^{-\delta/\varepsilon}).$$

Thus the unique solution of (3.12) is of the form

$$y(x,\varepsilon) = \tilde{A}(x,\varepsilon) + \tilde{B}(x,\varepsilon)\exp\left[-\frac{1}{\varepsilon}\int_0^x a(s)\,ds\right], \qquad (3.13)$$

where

$$\tilde{A}(x,\varepsilon) = \beta(\varepsilon)\frac{A(x,\varepsilon)}{A(1,\varepsilon)} + O(e^{-\delta/\varepsilon})$$

and

$$\tilde{B}(x,\varepsilon) = \left(\alpha(\varepsilon) - \frac{\beta(\varepsilon)}{A(1,\varepsilon)}\right)B(x,\varepsilon) + O(e^{-\delta/\varepsilon}).$$

Further, since

$$\tilde{A}(x,0) = \beta(0)\exp\left[-\int_1^x \frac{b(s)}{a(s)}\,ds\right],$$

the limiting solution for $x > 0$ satisfies the reduced problem

$$a(x)y' + b(x)y = 0, \qquad 0 \leq x \leq 1$$
$$y(1) = \beta(0).$$

Moreover, $\tilde{A}(0,0) \neq \alpha(0)$, in general, so we must expect nonuniform convergence at $x = 0$. Crudely, then, we cancel the boundary condition at $x = 0$ and integrate the resulting terminal value problem for the reduced equation to get the limiting solution for (3.12). Before going on, we note that a direct (but more sophisticated) proof of the asymptotic validity of (3.13) has been given by Cochran (1962) [see also O'Malley (1969a)]. Our results are pictured in Figs. 7 and 8.

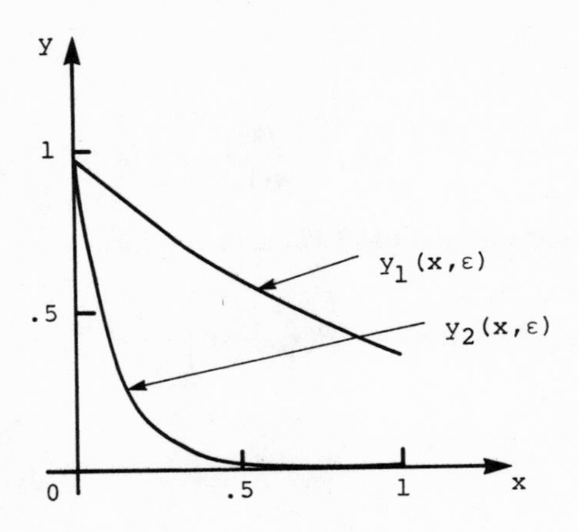

FIGURE 7 The linearly independent solutions y_1 and y_2 of $\varepsilon y'' + y' + y = 0$, $\varepsilon = 0.1$.

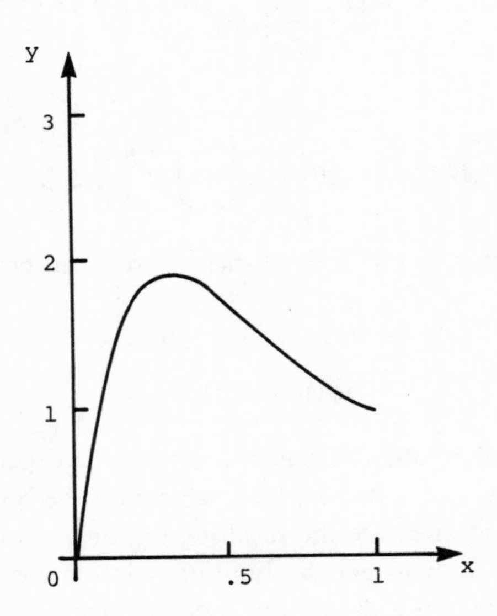

FIGURE 8 The solution of $\varepsilon y'' + y' + y = 0$, $y(0) = 0$, $y(1) = 1$, ε small.

Analogously, we find that, for the initial value problem

$$\begin{aligned} &\varepsilon y'' + a(x)y' + b(x)y = 0, \qquad 0 \le x \le 1 \\ &y(0) = \alpha(\varepsilon), \qquad y'(0) = \gamma(\varepsilon), \end{aligned} \tag{3.14}$$

we can also determine the coefficients k_1 and k_2 uniquely and we have

$$\begin{aligned} y(x,\varepsilon) = &\left[\frac{a(0)\alpha + \varepsilon(\gamma - \alpha B'(0,\varepsilon))}{a(0) + \varepsilon(A'(0,\varepsilon) - B'(0,\varepsilon))}\right] A(x,\varepsilon) \\ &+ \varepsilon\left[\frac{\alpha A'(0,\varepsilon) - \gamma}{a(0) + \varepsilon(A'(0,\varepsilon) - B'(0,\varepsilon))}\right] B(x,\varepsilon) \\ &\times \exp\left[-\frac{1}{\varepsilon}\int_0^x a(s)\,ds\right] \end{aligned} \tag{3.15}$$

up to asymptotically exponentially small terms. Here, then, the limiting solution $\alpha(0)\,a_0(x)$ for $0 \le x \le 1$ satisfies the reduced problem

$$\begin{aligned} a(x)y' + b(x)y &= 0 \\ y(0) &= \alpha(0); \end{aligned}$$

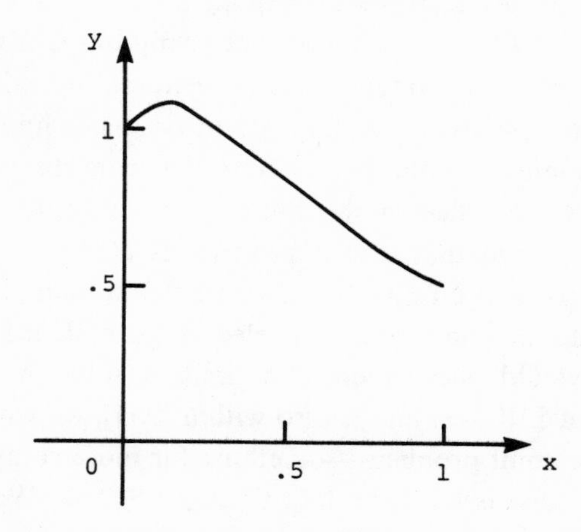

FIGURE 9 The solution of $\varepsilon y'' + y' + y = 0$, $y(0) = 1$, $y'(0) = 2$, $\varepsilon = 0.1$.

i.e., the derivative boundary condition is canceled in defining the appropriate limiting problem. Thus, $y(x, \varepsilon)$ will converge uniformly on $[0, 1]$ while $y'(x, \varepsilon)$ will converge nonuniformly at $x = 0$, and higher derivatives will be initially unbounded as $\varepsilon \to 0$. (See Fig. 9.)

Finally, for the terminal value problem

$$\begin{aligned} \varepsilon y'' + a(x)y' + b(x)y &= 0, & 0 \le x \le 1 \\ y(1) &= \beta(\varepsilon), & y'(1) = \kappa(\varepsilon), \end{aligned} \tag{3.16}$$

we can again determine the coefficients k_1 and k_2 of (3.11) uniquely. Substituting, then, we obtain

$$\begin{aligned} y(x, \varepsilon) = &\Big\{ [\beta a(1) B(1, \varepsilon) + \varepsilon\big(\kappa B(1, \varepsilon) - \beta B'(1, \varepsilon)\big)] A(x, \varepsilon) \\ &+ \frac{1}{\varepsilon}[\beta A'(1, \varepsilon) - \kappa A(1, \varepsilon)] B(x, \varepsilon) \exp\left[\frac{1}{\varepsilon}\int_x^1 a(s)\, ds\right]\Big\} \\ &\times \{a(1) A(1, \varepsilon) B(1, \varepsilon) + \varepsilon\big(B(1, \varepsilon) A'(1, \varepsilon) - A(1, \varepsilon) B'(1, \varepsilon)\big)\}^{-1}. \end{aligned} \tag{3.17}$$

This solution, however, blows up exponentially for $x < 1$ as $\varepsilon \to 0$ [unless perhaps $\beta A'(1, \varepsilon) = \kappa A(1, \varepsilon)$] since $\exp[(1/\varepsilon) \int_x^1 a(s)\, ds]$ does so.

It seems from these examples that, whenever the solution of Eq. (3.1) subject to separated boundary conditions has a limiting solution as $\varepsilon \to 0$, a boundary condition at $x = 0$ is canceled to define the reduced problem satisfied by this limiting solution for $x > 0$. This is because we assumed that $a(x)$ is positive. If $a(x)$ were negative on $[0, 1]$, the transformation $z = 1 - x$ shows that a boundary condition at $x = 1$ would then have to be canceled, in general, and nonuniform convergence would occur there if a limiting solution within $[0, 1]$ could be defined. If $a(x)$ had a zero within $[0, 1]$, we would be faced with a turning point problem—something far more complicated than the examples discussed here [cf. Chapter 8 and Wasow (1968), however].

2. HIGHER-ORDER PROBLEMS

Following Wasow (1941, 1944), we will consider the boundary value problem consisting of the linear differential equation

$$\varepsilon^{m-n}\left(y^{(m)} + \alpha_1(x)y^{(m-1)} + \cdots + \alpha_m(x)y\right) + \beta(x)\left(y^{(n)} + \beta_1(x)y^{(n-1)} + \cdots + \beta_n(x)y\right) = 0 \tag{3.18}$$

on the interval $0 \leq x \leq 1$ and the boundary conditions

$$\begin{aligned} y^{(\lambda_i)}(0) &= l_i, \qquad i = 1, 2, \cdots, r \\ y^{(\lambda_i)}(1) &= l_i, \qquad i = r+1, \cdots, m. \end{aligned} \tag{3.19}$$

We shall take $m > n$ so that the order of the differential equation drops when $\varepsilon = 0$, and shall assume that the coefficients of (3.18) are real and infinitely differentiable throughout $[0, 1]$ and that

$$\beta(x) \neq 0 \tag{3.20}$$

there. (The last assumption eliminates the possibility that the limiting nth-order equation is singular within $[0, 1]$.) Further, we will order the boundary conditions so that

$$m > \lambda_1 > \lambda_2 > \cdots > \lambda_r \geq 0 \tag{3.21a}$$

and

$$m > \lambda_{r+1} > \lambda_{r+2} > \cdots > \lambda_m \geq 0, \tag{3.21b}$$

and we shall seek asymptotic solutions as the small positive parameter $\varepsilon \to 0$. (We introduce the parameter as ε^{m-n} for later convenience. In physical problems, one would naturally expect more general parameter dependence. Analysis should then proceed analogously, however.)

Since the differential equation is linear, any solution will be a linear combination of any set of m linearly independent solutions. Thus, we shall first determine m linearly independent asymptotic solutions of the form

$$Y_j(x, \varepsilon) = G_j(x, \varepsilon) \exp\left[\frac{1}{\varepsilon}\int_{x_j}^{x} \mu_j(t)\, dt\right], \qquad j = 1, 2, \cdots, m, \quad (3.22)$$

where x_j is either 0 or 1 and where each G_j has an asymptotic series expansion

$$G_j(x, \varepsilon) \sim \sum_{i=0}^{\infty} G_{ji}(x)\varepsilon^i$$

as $\varepsilon \to 0$ with coefficients that are infinitely differentiable functions of x such that $G_{j0}(x) \not\equiv 0$. Differentiating k times, we have

$$\begin{aligned} Y_j^{(k)}(x, \varepsilon) = &\left\{G_j^{(k)} + \frac{k}{\varepsilon} G_j^{(k-1)} \mu_j + \cdots + \frac{1}{\varepsilon^{k-1}}\right. \\ &\times \left.\left(kG_j' \mu_j^{k-1} + \frac{k(k-1)}{2} G_j \mu_j^{k-2} \mu_j'\right) + \frac{G_j \mu_j^k}{\varepsilon^k}\right\} \\ &\times \exp\left[\frac{1}{\varepsilon}\int_{x_j}^{x} \mu_j(t)\, dt\right], \end{aligned} \quad (3.23)$$

so substituting into the differential equation (3.18) and collecting terms we obtain

$$\begin{aligned} \frac{1}{\varepsilon^n}&\left\{G_j\left(\mu_j^m + \beta(x)\mu_j^n\right)\right. \\ &+ \varepsilon\left[G_j'\left(m\mu_j^{m-1} + n\,\beta(x)\mu_j^{n-1}\right)\right. \\ &+ G_j\left(\frac{m(m-1)}{2}\mu_j^{m-2}\mu_j' + \frac{n(n-1)}{2}\beta(x)\mu_j^{n-2}\mu_j'\right. \\ &+ \left.\left.\alpha_1(x)\mu_j^{m-1} + \beta_1(x)\,\beta(x)\mu_j^{n-1}\right)\right] + \varepsilon^2(\cdots) + \varepsilon^n \beta(x) \\ &\times \left.\left[G_j^{(n)} + \beta_1(x)\, G_j^{(n-1)} + \cdots + \beta_n(x) G_j\right]\right\} \\ &\times \exp\left[\frac{1}{\varepsilon}\int_{x_j}^{x} \mu_j(t)\, dt\right] = 0. \end{aligned} \quad (3.24)$$

Thus, the leading term of (3.24) implies that we should ask that

$$\mu_j^m + \beta(x)\mu_j^n = 0, \tag{3.25}$$

obtaining $m - n$ distinct roots $\mu_j(x) = (-\beta(x))^{1/(m-n)}$ which are nowhere zero in $[0, 1]$ and n roots which are identically zero. Let these roots be ordered so that

$$\operatorname{Re} \mu_j(x) < 0 \qquad \text{in } [0, 1] \qquad \text{for} \quad j = 1, 2, \ldots, \sigma \tag{3.26a}$$

$$\operatorname{Re} \mu_j(x) > 0 \qquad \text{for} \quad j = m - n - \tau + 1, \ldots, m - n \tag{3.26b}$$

and

$$\mu_j(x) \equiv 0 \qquad \text{for} \quad j = m - n + 1, \ldots, m. \tag{3.26c}$$

Two cases arise: the nonexceptional case when $\sigma + \tau = m - n$ (i.e., when $m - n$ μ_j's have nonzero real parts), and the exceptional case when $\sigma + \tau = m - n - 2$ (i.e., $\mu_{\sigma+1} = -\mu_{\sigma+2} \neq 0$ are imaginary), since the roots of (3.25) are distributed like the $(m - n)$th roots of either ± 1. [Determining the sign of the real parts of the μ_j's is clearly related to stability considerations. As such, analogous calculations were important in early control theory work (cf. Meerov, 1961) and in thin shell theory (cf. Gol'denveizer, 1961).]

The coefficients G_{jk} are determined successively so that higher-order coefficients of ε^j in (3.24) become zero. Thus, for $j \leq m - n$, setting the coefficient of ε^{1-n} to zero implies that the G_{j0} must satisfy

$$G_{j0}'\left(m\mu_j^{m-1} + n\,\beta(x)\,\mu_j^{n-1}\right) + G_{j0}\left(\frac{m(m-1)}{2}\mu_j^{m-2}\mu_j' + \cdots + \beta_1(x)\,\beta(x)\mu_j^{n-1}\right) = 0. \tag{3.27}$$

Noting that $m\mu_j^{m-1} + n\,\beta(x)\,\mu_j^{n-1} \neq 0$, because the nonzero roots of

(3.25) are distinct,

$$G_{j0} = \frac{g_{j0} \exp\left[\int^x \frac{(\beta_1(s) - \alpha_1(s))}{(m-n)} ds\right]}{|\beta(x)|^{(m-n+1)/2(m-n)}},$$

where $g_{j0} \neq 0$ is undetermined. Successive G_{jk}'s satisfy nonhomogeneous forms of (3.27) and each is uniquely determined up to an arbitrary constant. Without later loss of generality, then, we will take $G_j(x_j, \varepsilon) = 1$ for $j \leq m - n$.

For $j > m - n$, $\mu_j \equiv 0$ so $G_j(x, \varepsilon)$ must satisfy the original ("full") differential equation (3.18). Equating coefficients when $\varepsilon = 0$ implies that G_{j0} must satisfy the reduced equation

$$y^{(n)} + \beta_1(x)y^{(n-1)} + \cdots + \beta_n(x)y = 0. \tag{3.28}$$

Likewise, successive G_{ji}'s for $i > 0$ must satisfy nonhomogeneous forms of (3.28). Corresponding to the n linearly independent solutions of (3.28), then, we formally obtain n linearly independent asymptotic solutions of (3.18).

It is worth noting that the procedure used to obtain the m formal solutions $Y_j(x, \varepsilon)$ of the form (3.22) is completely analogous to that used in the geometrical theory of optics [cf. Keller and Lewis (to be published)]. Equation (3.25) corresponds to the eiconal equations and (3.27) to the transport equations of that theory.

That the m asymptotic solutions formally constructed form a fundamental system (as $\varepsilon \to 0$) follows from the results of Turrittin (1936). His analysis, as might be expected, is based on integral equations. It also shows that these asymptotic solutions may be formally differentiated termwise.

Under appropriate conditions, the solution of the boundary value problem (3.18)–(3.19) will converge within (0, 1) to the solution of a reduced boundary value problem as $\varepsilon \to 0$. The reduced problem will consist of the reduced equation [i.e., (3.18) with $\varepsilon = 0$] and n of the m boundary conditions (3.19). We tell which $m - n$ boundary conditions are omitted in defining the reduced problem by applying a cancellation law. Before stating the somewhat complicated law, we note that it could be motivated by considering a series of simple examples with easily determined limiting behavior. We encourage the reader to examine such problems for himself and merely state:

THE CANCELLATION LAW

(1) Cancel σ boundary conditions at $x = 0$ and τ boundary conditions at $x = 1$, starting from those with the highest derivatives (i.e., largest λ_j's).

(2) In the exceptional case when $\sigma + \tau = m - n - 2$ also cancel from the remaining boundary conditions those two with the highest order λ_j of differentiation, requiring that they belong to the same endpoint, say $\tilde{x}$, and that their selection must be without ambiguity.

Let S and T be the total number of boundary conditions canceled at $x = 0$ and $x = 1$, respectively. (Thus, $S = \sigma$ and $T = \tau$ when $\sigma + \tau = m - n$. Otherwise, $S = \sigma + 2$ and $T = \tau$ when $\tilde{x} = 0$ and $S = \sigma$ and $T = \tau + 2$ when $\tilde{x} = 1$.) The cancellation law is well defined if and only if

$$S \leq r \qquad \text{and} \qquad T \leq m - r \tag{3.29}$$

and, in addition, in the exceptional case,

$$\lambda_S > \lambda_{r+T+1} \qquad \text{if} \quad m - r > T \quad \text{and} \quad \tilde{x} = 0 \tag{3.30a}$$

or

$$\lambda_{r+T} > \lambda_{S+1} \qquad \text{if} \quad r > S \quad \text{and} \quad \tilde{x} = 1. \tag{3.30b}$$

Under these conditions, the reduced problem is well defined as

$$\begin{aligned} &z^{(n)} + \beta_1(x) z^{(n-1)} + \cdots + \beta_n(x) z = 0, \qquad 0 \leq x \leq 1 \\ &z^{(\lambda_i)}(0) = l_i, \qquad i = S + 1, \ldots, r \\ &z^{(\lambda_i)}(1) = l_i, \qquad i = r + T + 1, \ldots, m. \end{aligned} \tag{3.31}$$

NOTE: If $m - n$ is odd, (3.25) implies that the nonexceptional case holds and $|S - T| = |\sigma - \tau| = 1$. If $m - n$ is even and $\sigma + \tau = m - n$, then $\sigma = S = T = \tau$, while if $m - n$ is even and $\sigma + \tau = m - n - 2$, then $\sigma = \tau$ but $|S - T| = 2$. In all cases, then, the total number of boundary conditions canceled at each endpoint differ by two or less. (For more general problems than (3.18)–(3.19), this is no longer true [cf. O'Malley and Keller (1968)].)

Since the $G_{m-n+k}(x, 0)$, $k = 1, 2, \cdots, n$, form a fundamental set of solutions for the reduced equation (3.28), the reduced problem will have a unique solution provided the $n \times n$ matrix

$$\phi = \begin{bmatrix} G_{m-n+k}^{(\lambda_{S+j})}(0,0) \\ G_{m-n+k}^{(\lambda_{r+T+l})}(1,0) \end{bmatrix}_{j=1,\ldots,r-S,\ l=1,\ldots,m-r-T, k=1,\ldots,n} \tag{3.32}$$

is nonsingular [cf., e.g., Coddington and Levinson (1955)]. We are now ready to state our theorem. Before doing so, let us define the integers γ_0 and γ_1:

$$\gamma_0 = \lambda_S \quad \text{and} \quad \gamma_1 = \lambda_{r+T} \quad \text{if} \quad \sigma + \tau = m - n. \tag{3.33a}$$

Otherwise

$$\gamma_0 = \min(\lambda_S, \lambda_{r+T}) \quad \text{and} \quad \gamma_1 = \lambda_{r+T} \quad \text{if} \quad \tilde{x} = 1 \tag{3.33b}$$

and

$$\gamma_0 = \lambda_S \quad \text{and} \quad \gamma_1 = \min(\lambda_S, \lambda_{r+T}) \quad \text{if} \quad \tilde{x} = 0. \tag{3.33c}$$

Then, recalling that $p = q$ (modulo $m - n$) if $m - n$ is a factor of $p - q$, we have

THEOREM 2: *Consider the boundary value problem* (3.18)–(3.19), *where the reduced problem* (3.31) *is well defined and has a unique solution* $z(x)$. *Assume further that:*

(a) $\lambda_1, \lambda_2, \ldots, \lambda_S$ *are distinct modulo* $m - n$, and

(b) $\lambda_{r+1}, \lambda_{r+2}, \ldots, \lambda_{r+T}$ *are distinct modulo* $m - n$. *Then, for* ε *sufficiently small, the problem* (3.18)–(3.19) *has a unique solution. It has the form*

$$y(x,\varepsilon) = \tilde{G}(x,\varepsilon) + \varepsilon^{\gamma_0} \sum_{k=1}^{S} \tilde{G}_k(x,\varepsilon) \exp\left[\frac{1}{\varepsilon}\int_0^x \mu_j(t)\,dt\right]$$

$$+ \varepsilon^{\gamma_1} \sum_{k=S+1}^{m-n} \tilde{G}_k(x,\varepsilon) \exp\left[\frac{1}{\varepsilon}\int_1^x \mu_j(t)\,dt\right], \qquad (3.34)$$

where the functions $\tilde{G}$ and $\tilde{G}_k$ have asymptotic power series expansions in ε for $0 \le x \le 1$ with coefficients which may be obtained successively by a scheme of undetermined coefficients. Further,

$$z(x) = \tilde{G}(x,0). \qquad (3.35)$$

Thus,

$$y(x,\varepsilon) \to z(x) \quad \text{as} \quad \varepsilon \to 0$$

in the open interval $0 < x < 1$ (but for the exceptional cases when either $\tilde{x} = 0$ and $\gamma_0 = 0$ or $\tilde{x} = 1$ and $\gamma_1 = 0$ when the limiting solution oscillates rapidly about z).

We note that the representation (3.34) can be formally differentiated repeatedly. In particular, then, in the nonexceptional case, the exponential factors are asymptotically negligible within $0 < x < 1$ and we have

$$\frac{dy^{(j)}(x,\varepsilon)}{dx^j} \to z^{(j)}(x)$$

there as $\varepsilon \to 0$.

The theorem gives sufficient conditions for convergence to a limiting solution. That the hypotheses are nearly necessary for the problem is shown by the following list of constant coefficient examples whose solutions diverge as $\varepsilon \to 0$.

1. $$\varepsilon^3 y^{(4)} + y' = 0$$

$$y''(0) = y'(0) = y(0) = 0, \qquad y(1) = 1.$$

Here $\tau = 2$, so we would need to cancel two boundary conditions at $x = 1$. Since only one boundary condition is given there, however, the reduced problem (cancellation law) is not well defined.

2.
$$\varepsilon y^{(2)} - y' = 0$$
$$y'(0) = 1, \qquad y(1) = 0.$$

Here the reduced problem $z' = 0$, $z'(0) = 1$ has no solution.

3.
$$\varepsilon^3 y^{(4)} - y' = 0$$
$$y'''(0) = y(0) = y(1) = 0, \qquad y'(1) = 1.$$

Here the two boundary conditions to be canceled at $x = 0$ have orders of differentiation differing by $m - n = 3$.

4.
$$\varepsilon^2 y^{(3)} + y' = 0$$
$$y'(0) = 0, \qquad y(0) = y'(1) = 1.$$

Here the exceptional case holds, but the reduced problem is not well defined since the last two boundary conditions to be canceled belong to different endpoints.

5.
$$\varepsilon^2 y^{(4)} + y^{(2)} = 0$$
$$y''(0) = y'(0) = y'(1) = 0, \qquad y(1) = 1.$$

Here, the boundary conditions to be canceled are not well determined.

A SAMPLE PROBLEM: Let us consider the boundary value problem

$$\varepsilon^2 y^{(4)} - y'' = 0, \qquad 0 \leq x \leq 1$$

with

$$y'''(0), \quad y(0), \quad y'(1), \quad \text{and} \quad y(1)$$

being prescribed constants. Here the exponents μ_j are roots of the polynomial $\mu^4 - \mu^2 = 0$ so $\sigma = \tau = S = T = 1$ and Theorem 2 implies that there will be a unique asymptotic solution which is of the form

$$y(x, \varepsilon) = Z(x, \varepsilon) + \varepsilon^3 L(x, \varepsilon)\, e^{-x/\varepsilon} + \varepsilon R(x, \varepsilon)\, e^{-(1-x)/\varepsilon},$$

where Z, L, and R have power series expansions in ε and $Z(x,0)$ is the unique solution of the reduced problem

$$Z'' = 0$$
$$Z(0) = y(0), \qquad Z(1) = y(1).$$

We will show how the first few terms are determined by using undetermined coefficients. The reader should be aware that such an expansion procedure is much more efficient than obtaining a fundamental set of asymptotic approximations (as in the preceding section) and then seeking the correct linear combination to satisfy the boundary conditions.

Setting

$$\begin{aligned} y(x,\varepsilon) = {} & Z_0(x) + \varepsilon[Z_1(x) + R_0(x)e^{-(1-x)/\varepsilon}] \\ & + \varepsilon^2[Z_2(x) + R_1(x)e^{-(1-x)/\varepsilon}] \\ & + \varepsilon^3[Z_3(x) + R_2(x)e^{-(1-x)/\varepsilon} + L_0(x)e^{-x/\varepsilon}] + O(\varepsilon^4) \end{aligned}$$

and differentiating formally, we have

$$\begin{aligned} y'(x,\varepsilon) = {} & (Z_0' + R_0 e^{-(1-x)/\varepsilon}) + \varepsilon\left(Z_1' + (R_1 + R_0')e^{-(1-x)/\varepsilon}\right) \\ & + \varepsilon^2\left(Z_2' + (R_2 + R_1')e^{-(1-x)/\varepsilon} - L_0 e^{-x/\varepsilon}\right) + O(\varepsilon^3) \end{aligned}$$

$$\begin{aligned} y''(x,\varepsilon) = {} & \frac{R_0}{\varepsilon}e^{-(1-x)/\varepsilon} + \left(Z_0'' + (R_1 + 2R_0')e^{-(1-x)/\varepsilon}\right) \\ & + \varepsilon[Z_1'' + (R_2 + 2R_1' + R_0'')e^{-(1-x)/\varepsilon} + L_0 e^{-x/\varepsilon}] + O(\varepsilon^2) \end{aligned}$$

$$\begin{aligned} y'''(x,\varepsilon) = {} & \frac{R_0}{\varepsilon^2}e^{-(1-x)/\varepsilon} + \frac{1}{\varepsilon}(R_1 + 3R_0')e^{-(1-x)/\varepsilon} \\ & + \left(Z_0''' + (R_2 + 3R_1' + 3R_0'')e^{-(1-x)/\varepsilon} - L_0 e^{-x/\varepsilon}\right) + O(\varepsilon) \end{aligned}$$

$$\begin{aligned} y^{\text{iv}}(x,\varepsilon) = {} & \frac{R_0}{\varepsilon^3}e^{-(1-x)/\varepsilon} + \frac{1}{\varepsilon^2}(R_1 + 4R_0')e^{-(1-x)/\varepsilon} \\ & + \frac{1}{\varepsilon}[(R_2 + 4R_1' + 6R_0'')e^{-(1-x)/\varepsilon} + L_0 e^{-x/\varepsilon}] + O(1). \end{aligned}$$

Thus, the differential equation becomes

$$[-Z_0'' + 2R_0' e^{-(1-x)/\varepsilon}] + \varepsilon[-Z_1'' + (2R_1' + 5R_0'')e^{-(1-x)/\varepsilon}] + O(\varepsilon^2)$$
$$= 0.$$

This implies that

$$Z_0'' = 0$$
$$R_0' = 0$$

and

$$Z_1'' = 0$$

among other equations. Further, the boundary conditions become

$$y(0) = Z_0(0) + \varepsilon Z_1(0) + \varepsilon^2 Z_2(0) + \varepsilon^3\big(Z_3(0) + L_0(0)\big) + O(\varepsilon^4)$$
$$y(1) = Z_0(1) + \varepsilon\big(Z_1(1) + R_0(1)\big) + \varepsilon^2\big(Z_2(1) + R_1(1)\big)$$
$$+ \varepsilon^3\big(Z_3(1) + R_2(1)\big) + O(\varepsilon^4)$$
$$y'(1) = \big(Z_0'(1) + R_0(1)\big) + \varepsilon\big(Z_1'(1) + R_1(1) + R_0'(1)\big)$$
$$+ \varepsilon^2\big(Z_2'(1) + R_2(1) + R_1'(1)\big) + O(\varepsilon^3)$$
$$y'''(0) = \big(Z_0'''(0) - L_0(0)\big) + O(\varepsilon).$$

Thus, we have

$$Z_0(0) = y(0)$$
$$Z_1(0) = 0$$
$$Z_0(1) = y(1)$$
$$Z_1(1) + R_0(1) = 0$$

and

$$Z_0'(1) + R_0(1) = y'(1)$$

among other equations. As expected, then,

$$Z_0'' = 0$$
$$Z_0(0) = y(0), \qquad Z_0(1) = y(1)$$

or

$$Z_0(x) = y(0) + \big(y(1) - y(0)\big)x.$$

Further,

$$R_0'(x) = 0$$
$$R_0(1) = y'(1) - Z_0'(1)$$

implies that

$$R_0(x) = y'(1) - y(1) + y(0).$$

Finally,

$$Z_1''(x) = 0$$
$$Z_1(0) = 0, \qquad Z_1(1) = -R_0(1)$$

so that

$$Z_1(x) = \big(-y'(1) + y(1) - y(0)\big)x.$$

Thus,

$$\begin{aligned} y(x, \varepsilon) = {} & [y(0) + \big(y(1) - y(0)\big)x] \\ & + \varepsilon\Big[\big(-y'(1) + y(1) - y(0)\big)(x - e^{-(1-x)/\varepsilon})\Big] \\ & + O(\varepsilon^2) \end{aligned}$$

and

$$\begin{aligned} y'(x, \varepsilon) = {} & [y(1) - y(0)] - \big(-y'(1) + y(1) - y(0)\big)e^{-(1-x)/\varepsilon} \\ & + O(\varepsilon). \end{aligned}$$

In this example, then, $y(x, \varepsilon)$ converges to the solution $Z_0(x)$ of the reduced problem as $\varepsilon \to 0$ throughout $[0, 1]$, while $y'(x, \varepsilon)$ converges to $Z_0'(x)$ except near $x = 1$, where nonuniform convergence occurs [unless $y(0) = y(1) - y'(1)$]. Higher-order derivatives of $Z_0(x)$ are zero, and higher-order derivatives of $y(x, \varepsilon)$ converge to zero except at $x = 0$ and $x = 1$, where they will generally be (algebraically) unbounded as $\varepsilon \to 0$.

A special example is illustrated in Fig. 10. Further examples may be found in the book by B. Friedman (1969).

PROOF OF THEOREM 2: Consider the nonexceptional case where we can write the general solution of the differential equation

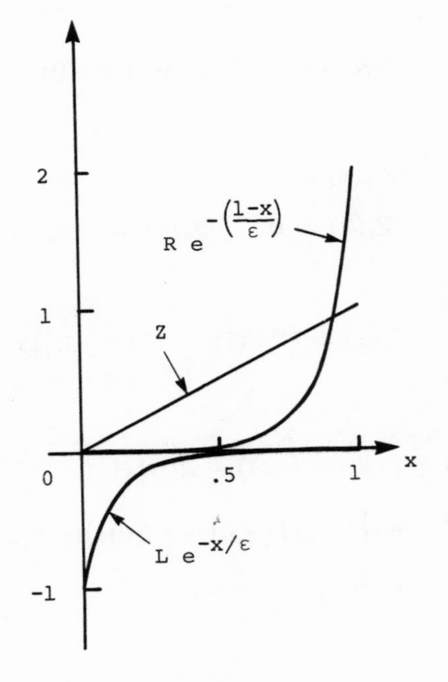

FIGURE 10 The component parts $Z(x, \varepsilon)$, $L(x, \varepsilon)e^{-x/\varepsilon}$, and $R(x, \varepsilon)e^{-(1-x)/\varepsilon}$ of the solution $y(x, \varepsilon) = Z(x, \varepsilon) + \varepsilon^3 L(x, \varepsilon)e^{-x/\varepsilon} + \varepsilon R(x, \varepsilon)e^{-(1-x)/\varepsilon}$ to $\varepsilon^2 y^{(4)} - y^{(2)} = 0$, $y(0) = 0$, $y'''(0) = 1$, $y(1) = 1$, $y'(1) = 3$.

(3.18) as

$$y(x, \varepsilon) = \varepsilon^{\lambda_\sigma} \sum_{j=1}^{\sigma} C_j(\varepsilon) G_j(x, \varepsilon) \exp\left[\frac{1}{\varepsilon}\int_0^x \mu_j(t)\,dt\right]$$
$$+ \varepsilon^{\lambda_{r+\tau}} \sum_{j=\sigma+1}^{m-n} C_j(\varepsilon) G_j(x, \varepsilon) \exp\left[\frac{1}{\varepsilon}\int_1^x \mu_j(t)\,dt\right]$$
$$+ \sum_{j=m-n+1}^{m} C_j(\varepsilon) G_j(x, \varepsilon)$$

for undetermined coefficients $C_j(\varepsilon)$ [cf. Eq. (3.22)]. Using the differentiation formula (3.23), the boundary conditions $y^{(\lambda_i)}(0) = l_i$ take the form

$$l_i = \varepsilon^{\lambda_\sigma - \lambda_i} \sum_{j=1}^{\sigma} C_j(\varepsilon)\left[\left(\mu_j(0)\right)^{\lambda_i} + O(\varepsilon)\right]$$
$$+ \sum_{j=\sigma+1}^{m-n} C_j(\varepsilon) \qquad \text{(asymptotically exponentially small terms)}$$
$$+ \sum_{j=m-n+1}^{m} C_j(\varepsilon)\left[G_j^{(\lambda_i)}(0,0) + O(\varepsilon)\right], \qquad \text{for} \quad i = 1, 2, \ldots, r$$

while $y^{(\lambda_i)}(1) = l_i$ imply

$$l_i = \sum_{j=1}^{\sigma} C_j(\varepsilon) \quad \text{(asymptotically exponentially small terms)}$$
$$+ \varepsilon^{\lambda_{r+\tau} - \lambda_i} \sum_{j=\sigma+1}^{m-n} C_j(\varepsilon)\left[\left(\mu_j(1)\right)^{\lambda_i} + O(\varepsilon)\right]$$
$$+ \sum_{j=m-n+1}^{m} C_j(\varepsilon)\left[G_j^{(\lambda_i)}(1,0) + O(\varepsilon)\right], \qquad \text{for} \quad i = r+1, \ldots, m,$$

since we can take $G_j(x_j, 0) = 1, j \leq m - n$, with $x_j = 0$ for $j = 1, 2, \ldots, \sigma$ and $x_j = 1$ for $j = \sigma + 1, \ldots, m - n$. Considering these as m linear equations in the m unknowns $C_j(\varepsilon)$, we can obtain a unique solution if the determinant of coefficients is nonzero. This determinant, however, is a nonzero multiple of

$$\Delta(\varepsilon) = \det\begin{bmatrix} \xi & X \\ 0 & \phi \end{bmatrix} + O(\varepsilon),$$

where X is a matrix with bounded entries, ϕ is given by (3.32), and

$$\xi = \begin{bmatrix} \xi_0 & 0 \\ 0 & \xi_1 \end{bmatrix}$$

for

$$\xi_0 = \left((\mu_j(0))^{\lambda_i}\right), \qquad i, j = 1, \ldots, \sigma$$

and

$$\xi_1 = \left((\mu_{\sigma+j}(1))^{\lambda_{r+i}}\right), \qquad i, j = 1, \ldots, \tau.$$

Thus

$$\Delta(0) = (\det \phi)(\det \xi_0)(\det \xi_1).$$

Note, first, that $\det \phi \neq 0$ [cf. (3.32)] since the reduced problem has a unique solution. Next, note that $\mu_1(0), \mu_2(0), \ldots, \mu_\sigma(0)$ are those determinations of $(-\beta(0))^{1/(m-n)}$ in the left-half plane. Thus, they are nonzero constant multiples of $\kappa, \kappa^2, \ldots, \kappa^\sigma$, where $\kappa = e^{(2\pi i)/(m-n)}$. This implies that

$$\det \xi_0 = C \det\left((\kappa^{\lambda_i})^j\right)$$
$$i, j = 1, 2, \ldots, \sigma, \qquad \text{for some} \quad C \neq 0.$$

Thus, $\det \xi_0 = 0$ if and only if a Vandermonde determinant is zero; i.e., if and only if

$$\kappa^{\lambda_\alpha} = \left(e^{(2\pi i)/(m-n)}\right)^{\lambda_\alpha} = \left(e^{(2\pi i)/(m-n)}\right)^{\lambda_\beta} = \kappa^{\lambda_\beta}$$

or

$$\left(e^{(2\pi i)/(m-n)}\right)^{\lambda_\alpha - \lambda_\beta} = 1$$

for integers α and β such that $1 \leq \alpha < \beta \leq \sigma$. [For a discussion of Vandermonde determinants, see, e.g., Bellman (1970).] This equality is impossible here because $\lambda_1, \ldots, \lambda_\sigma$ are distinct modulo $m - n$. Thus $\det \xi_0 \neq 0$. Similarly, ξ_1 is nonsingular because $\lambda_{r+1}, \lambda_{r+2}, \ldots, \lambda_{r+\tau}$

are distinct modulo $m - n$. Thus, $\Delta(0) \neq 0$ and $\Delta(\varepsilon)$ will be nonzero for ε sufficiently small. It is easy to show that the $C_j(\varepsilon)$'s will have power series expansions as $\varepsilon \to 0$, which can be obtained by Cramer's rule. Alternatively, the expansion (3.34) could be obtained directly by an undetermined coefficients scheme [cf. O'Malley and Keller (1968) where a proof in the exceptional case is also given]. Finally, analogous results would also follow provided $\Delta(\varepsilon) \neq 0$ for $\varepsilon > 0$, even if $\Delta(0) = 0$. For example, it is simple to obtain the unique asymptotic solution of the form (3.34) to the problem

$$\varepsilon^2 y^{(4)} - y'' - \pi^2 y = 0$$

$$y(0) = y(1) = 0, \qquad y'(0) = y'(1) = 1$$

even though the solution of the corresponding reduced problem

$$Z'' + \pi^2 Z = 0$$

$$Z(0) = Z(1) = 0$$

is not unique [and, therefore, $\Delta(0) = 0$].

3. GENERALIZATIONS OF THESE RESULTS

Turrittin (1936), among others, shows how to construct asymptotic solutions for more general linear equations (without turning points). When the appropriate characteristic equation has multiple roots, the Puiseaux (Newton) polygon method must be used to obtain formal solutions. The proof of asymptotic validity uses integral equations. Analogously, Hukuhara (1937) and Turrittin (1952) show how to obtain asymptotic solutions for systems of linear differential equations whose order drops when $\varepsilon = 0$. Much more complicated situations can be analyzed [cf. Stengle (1971)]. Constructing asymptotic solutions for such linear equations is analogous to obtaining asymptotic solutions to polynomial equations with coefficients depending on a small parameter. Readers would find it instructive to study the polygon method for such problems [cf., e.g., Vainberg and Trenogin (1962) or Walker (1962)].

Knowing a fundamental set of asymptotic solutions, very general boundary value problems can be studied. Thus, O'Malley and Keller (1968) consider boundary value problems when the order of the boundary conditions changes when $\varepsilon = 0$; Harris (1960, 1973) considered boundary value problems for linear systems with coupled boundary conditions; Handelman *et al.* (1968) and Harris (1961) considered eigenvalue problems; O'Malley (1969c) considered nonhomogeneous equations; and O'Malley and Mazaika (1971) considered multipoint problems with discontinuous coefficients. Using spectral representation, these results can be applied to boundary value problems in Hilbert space [cf. A. Friedman (1969) and Bobisud and Calvert (1970)]. Direct applications of these results occur in many fields [see, e.g., Boyce and Handelman (1961) or Nau and Simmonds (1972) for a vibrations problem, Desoer and Shensa (1970) for a problem in electrical networks, and Keller (1973) for a nonlinear diffusion problem]. It is worth observing that cancellation laws can also be obtained for the frequently occurring case of equations with small coefficients multiplying the highest derivatives, but where explicit dependence on a small parameter is not known. Such results should be helpful in analyzing "stiff" differential equations [cf. Bjurel *et al.* (1970)]. A numerical technique for boundary value problems is discussed in Abrahamsson *et al.* (to appear).

A. The Vibrating String Problem

As a simple example, we shall consider the eigenvalue problem which arises from the vibrations of a string with clamped endpoints [see Lord Rayleigh (1945)]. If the string has negligible stiffness, the appropriate eigenvalue problem is

$$\begin{aligned} y'' &= -\lambda y, \qquad 0 \le x \le 1 \\ y(0) &= y(1) = 0. \end{aligned} \tag{3.36}$$

If stiffness effects are introduced, we must, instead, consider the higher-order problem

$$\begin{aligned} \varepsilon^2 y^{iv} - y'' &= \lambda y, \qquad 0 \le x \le 1 \\ y'(0) = y(0) &= y'(1) = y(1) = 0, \end{aligned} \tag{3.37}$$

where ε^2 is proportional to the stiffness. As the stiffness (and the positive parameter ε) tends to zero, we expect the eigenvalues and eigenfunctions of the full problem (3.37) to converge to those of the reduced problem (3.36), i.e., to $n^2\pi^2$ and $\sin n\pi x$, respectively, for $n = 1, 2, \ldots$. The eigenfunctions, however, might be expected to feature nonuniform convergence as $\varepsilon \to 0$ near both endpoints.

Four linearly independent asymptotic solutions of the differential equation $y'' + \lambda y = 0$; namely,

$$G_1(x,\lambda,\varepsilon)\, e^{-x/\varepsilon}, \quad G_2(x,\lambda,\varepsilon)\, e^{-(1-x)/\varepsilon},$$
$$G_3(x,\lambda,\varepsilon), \quad \text{and} \quad G_4(x,\lambda,\varepsilon) \tag{3.38}$$

can be formally obtained, as in Section 2. Here, the G_i's have asymptotic series expansions in ε with coefficients depending on x and λ and, without loss of generality, $G_1(0,\lambda,\varepsilon) = 1 = G_2(1,\lambda,\varepsilon)$. The general solution of the differential equation will be a linear combination of these four linearly independent solutions while the eigenvalues λ are those values such that the boundary value problem (3.37) has nontrivial solutions. The boundary conditions, then, imply a system of four linear equations for the coefficients of the linear combination. The system will have a nontrivial solution with λ as an eigenvalue if and only if the determinant of coefficients $\Delta(\lambda)$ vanishes. For ε sufficiently small, $\Delta(\lambda)$ has the form

$$\Delta(\lambda) = \frac{1}{\varepsilon^2}$$
$$\times \det \begin{pmatrix} 1 & 0 & G_3(0,\lambda,\varepsilon) & G_4(0,\lambda,\varepsilon) \\ 0 & 1 & G_3(1,\lambda,\varepsilon) & G_4(1,\lambda,\varepsilon) \\ -1 + \varepsilon G_1'(0,\lambda,\varepsilon) & 0 & \varepsilon G_3'(0,\lambda,\varepsilon) & \varepsilon G_4'(0,\lambda,\varepsilon) \\ 0 & 1 + \varepsilon G_2'(1,\lambda,\varepsilon) & \varepsilon G_3'(1,\lambda,\varepsilon) & \varepsilon G_4'(1,\lambda,\varepsilon) \end{pmatrix}$$
$$+ O\left(\frac{1}{\varepsilon^2} e^{-1/\varepsilon}\right). \tag{3.39}$$

From this it follows that the eigenvalues $\lambda(\varepsilon)$ as functions of ε all have asymptotic series expansions in ε with the leading terms λ_0 being roots of the equation

$$\delta_0(\lambda_0) = \det \begin{pmatrix} G_3(0,\lambda_0,0) & G_4(0,\lambda_0,0) \\ G_3(1,\lambda_0,0) & G_4(1,\lambda_0,0) \end{pmatrix} = 0. \tag{3.40}$$

Since $G_3(x,\lambda,0)$ and $G_4(x,\lambda,0)$ satisfy the reduced differential equation, it follows that λ_0 is an eigenvalue of the reduced problem (3.36). Note that if the eigenvalues $\lambda(\varepsilon)$ were known asymptotically, the corresponding eigenfunctions $y(x,\varepsilon)$ could be determined (uniquely up to normalization).

For the problem (3.37), then, it is natural to seek asymptotic expansions for the eigenvalues $\lambda(\varepsilon)$ and the corresponding eigenfunctions $y(x,\varepsilon)$ of the form

$$\lambda(\varepsilon) \sim \sum_{j=0}^{\infty} \lambda_j \varepsilon^j \tag{3.41}$$

and

$$y(x,\varepsilon) = A(x,\varepsilon) + B(x,\varepsilon)\,e^{-x/\varepsilon} + C(x,\varepsilon)\,e^{-(1-x)/\varepsilon}, \tag{3.42}$$

where A, B, and C have asymptotic series expansions in ε. We shall determine these expansions termwise by an undetermined coefficients procedure. Substituting into the differential equation and boundary conditions of (3.37), we find that the "outer expansion" $A(x,\varepsilon)$ must satisfy the full equation

$$\varepsilon^2 A^{\text{iv}} - A'' = \lambda A$$

while $B(x,\varepsilon)$ and $C(x,\varepsilon)$ must satisfy

$$\varepsilon^3 B^{\text{iv}} - 4\varepsilon^2 B''' + \varepsilon(5B'' - \lambda B) - 2B' = 0$$

$$\varepsilon^3 C^{\text{iv}} + 4\varepsilon^2 C''' + \varepsilon(5C'' - \lambda C) + 2C' = 0$$

and

$$0 = y(0,\varepsilon) \sim A(0,\varepsilon) + B(0,\varepsilon)$$

$$0 = y(1,\varepsilon) \sim A(1,\varepsilon) + C(1,\varepsilon)$$

$$0 = \varepsilon y'(0,\varepsilon) \sim \varepsilon\big(A'(0,\varepsilon) + B'(0,\varepsilon)\big) - B(0,\varepsilon)$$

$$0 = \varepsilon y'(1,\varepsilon) \sim \varepsilon\big(A'(1,\varepsilon) + C'(1,\varepsilon)\big) + C(1,\varepsilon).$$

Equating coefficients, then, we ask that

$$a''_j + \lambda_0 a_j = a^{\text{iv}}_{j-2} - \sum_{l=1}^{j} \lambda_l a_{j-l}$$
$$\equiv \begin{cases} 0 & \text{for} \quad j = 0 \\ -\lambda_j a_0 + \alpha_{j-1} & \text{for} \quad j \geq 1. \end{cases}$$

$$2b'_j = 5b''_{j-1} - \sum_{l=0}^{j-1} \lambda_l b_{j-1-l} - 4b'''_{j-2} + b^{\text{iv}}_{j-3} \equiv 2\beta_{j-1}$$

and

$$2c'_j = -5c''_{j-1} + \sum_{l=0}^{j-1} \lambda_l c_{j-1-l} - 4c'''_{j-2} - c^{\text{iv}}_{j-3} \equiv 2\gamma_{j-1}$$

and

$$a_j(0) = -b_j(0) = -a'_{j-1}(0) - b'_{j-1}(0)$$
$$a_j(1) = -c_j(1) = a'_{j-1}(1) + c'_{j-1}(1)$$

for all integers $j \geq 0$ where coefficients with negative subscripts are defined to be identical zero.

For $j = 0$, then

$$a''_0(x) + \lambda_0 a_0(x) = 0, \qquad a_0(0) = a_0(1) = 0$$
$$b'_0(x) = 0, \qquad b_0(0) = 0$$
$$c'_0(x) = 0, \qquad c_0(1) = 0.$$

Thus, $b_0(x) = c_0(x) \equiv 0$, and there is a nontrivial solution for a_0 only if $\lambda_0 = n^2\pi^2$, $n = 1, 2, \ldots$. Then

$$a_0(x) = A_0 \sin n\pi x$$

for arbitrary $A_0 \neq 0$. For a unique (up to sign) determination of a_0, we (somewhat arbitrarily) normalize by asking that $\int_0^1 a_0^2(s)\, ds = 1$ and set $A_0 = \sqrt{2}$.

For $j = 1$, we have

$$a_1''(x) + \lambda_0 a_1(x) = -\lambda_1 a_0(x), \qquad a_1(0) = -b_1(0)$$
$$a_1(1) = -c_1(1)$$
$$2b_1'(x) = 5b_0''(x) - \lambda_0 b_0(x), \qquad b_1(0) = a_0'(0)$$
$$2c_1'(x) = -5c_0''(x) + \lambda_0 c_0(x), \qquad c_1(1) = -a_0'(1).$$

Thus, $b_1(x) = \sqrt{2}\, n\pi$ and $c_1(x) = (-1)^n \sqrt{2}\, n\pi$ and, from the differential equation for a_0, we have

$$(a_0' a_1 - a_0 a_1')' = \lambda_1 a_0^2.$$

Integrating from 0 to 1, then $\lambda_1 = 4n^2\pi^2$, while the boundary value problem for a_1 implies that

$$a_1(x) = A_1 \sin n\pi x - \sqrt{2}\, n\pi(1 + 2x)\cos n\pi x$$

for A_1 arbitrary. To specify a_1 uniquely, we (again somewhat arbitrarily) impose the orthogonality condition

$$\int_0^1 a_0(s)\, a_1(s)\, ds = 0, \qquad \text{so} \quad A_1 = \sqrt{2}.$$

Continuing, suppose that the $a_l(x)$, $b_l(x)$, $c_l(x)$, and λ_l are known for $l < j$, $j \geq 1$. Then the differential equation and boundary conditions imply that

$$a_j''(x) + \lambda_0 a_j(x) = -\lambda_j a_0(x) + \alpha_{j-1}(x), \qquad a_j(0) = -b_j(0)$$
$$a_j(1) = -c_j(1)$$

$$b_j(x) = b_j(0) + \int_0^x \beta_{j-1}(s)\, ds$$

and

$$c_j(x) = c_j(1) + \int_1^x \gamma_{j-1}(s)\, ds,$$

where α_{j-1}, β_{j-1}, γ_{j-1}, $b_j(0)$, and $c_j(0)$ are known in terms of the preceding coefficients. Using the differential equations for a_0 and a_j, then

$$(a_0' a_j - a_0 a_j')' = \lambda_j a_0^2 - a_0 \alpha_{j-1},$$

so integration implies that

$$\lambda_j = b_j(0) a_0'(0) - c_j(1) a_0'(1) + \int_0^1 a_0(s) \alpha_{j-1}(s)\, ds.$$

The differential equation for a_j, then, yields $a_j(x)$ up to an arbitrary term $A_j \sin n\pi x$. The constant A_j can be uniquely specified by requiring that $\int_0^1 a_0(s) a_j(s)\, ds = 0$. Thus, all coefficients in the expansions for $\lambda(\varepsilon)$ and $y(x, \varepsilon)$ can be obtained termwise by this scheme of undetermined coefficients. That the resulting expansions are asymptotically correct has been proved by Handelman *et al.* (1968).

To summarize, we have shown that all eigenvalues of the vibrating string problem (3.37) with small stiffness are of the form

$$\lambda(\varepsilon) = n^2\pi^2 + 4\varepsilon n^2\pi^2 + \varepsilon^2(\cdots), \qquad n = 1, 2, \ldots \tag{3.43}$$

with corresponding (normalized) eigenfunctions

$$y(x, \varepsilon) = \sqrt{2}\, n\pi \left\{ \frac{\sin n\pi x}{n\pi} + \varepsilon \left[\frac{\sin n\pi x}{n\pi} - (1 - 2x) \cos n\pi x + e^{-x/\varepsilon} + (-1)^n e^{-(1-x)/\varepsilon} \right] + \varepsilon^2(\cdots) \right\}. \tag{3.44}$$

Note that the eigenvalues converge to $n^2\pi^2$ and the eigenfunctions to $\sqrt{2} \sin n\pi x$ as $\varepsilon \to 0$, as expected from the reduced problem (3.36). Convergence of the eigenfunctions is uniform throughout $0 \leq x \leq 1$, while derivatives of the eigenfunctions will converge nonuniformly as $\varepsilon \to 0$ at both endpoints $x = 0$ and $x = 1$. (This could have been anticipated, as in Theorem 2.)

B. Two-Parameter Problems

A direct and useful extension of the preceding results is to problems involving two parameters. Thus, we consider the boundary value problem consisting of the linear equation

$$\begin{aligned}&\varepsilon[y^{(m)} + \alpha_1(x)y^{(m-1)} + \cdots + \alpha_m(x)y] \\ &\quad + \mu\beta(x)[y^{(n)} + \beta_1(x)y^{(n-1)} + \cdots + \beta_n(x)y] \\ &\quad + \gamma(x)[y^{(p)} + \gamma_1(x)y^{(p-1)} + \cdots + \gamma_p(x)y] = 0 \end{aligned} \tag{3.45}$$

on the interval $0 \leq x \leq 1$ and the boundary conditions

$$\begin{aligned} y^{(\lambda_i)}(0) &= l_i, \qquad i = 1, 2, \ldots, r \\ y^{(\lambda_i)}(1) &= l_i, \qquad i = r+1, r+2, \ldots, m, \end{aligned} \tag{3.46}$$

where

$$m > \lambda_1 > \lambda_2 > \cdots > \lambda_r \geq 0$$

and

$$m > \lambda_{r+1} > \lambda_{r+2} > \cdots > \lambda_m \geq 0.$$

Here, ε and μ are small, positive parameters simultaneously approaching zero while

$$m > n > p \geq 0. \tag{3.47}$$

The limiting behavior of the asymptotic solutions to this differential equation will be completely different in the three cases where the parameters ε and μ are interrelated such that

(i) $\dfrac{\varepsilon}{\mu^{(m-p)/(n-p)}} \to 0,$

(ii) $\varepsilon = \mu^{(m-p)/(n-p)}$

or

(iii) $\dfrac{\mu^{(m-p)/(n-p)}}{\varepsilon} \to 0.$

These cases are all studied by O'Malley (1967b).

These sharp distinctions become already apparent in analyzing the simple constant coefficient equation

$$\varepsilon y'' + \mu a y' + by = 0, \qquad a \neq 0 \neq b,$$

where two linearly independent solutions are given by

$$y_1(x) = \exp\left[-\frac{\mu a x}{2\varepsilon}\left(1 + \left(1 - \frac{4\varepsilon b}{\mu^2 a^2}\right)^{1/2}\right)\right]$$

and

$$y_2(x) = \exp\left[-\frac{\mu a x}{2\varepsilon}\left(1 - \left(1 - \frac{4\varepsilon b}{\mu^2 a^2}\right)^{1/2}\right)\right].$$

When (i) $\varepsilon/\mu^2 \to 0$, note that the solutions behave approximately like $e^{-\mu a x/\varepsilon}$ and $e^{-bx/\mu a}$; when $\varepsilon = \mu^2$, they behave like $e^{-\mu a x/2\varepsilon} \times \exp[\mp(\mu a x/2\varepsilon)(1 - 4b/a^2)^{1/2}]$; and (iii), when $\mu^2/\varepsilon \to 0$, they are approximated by $\exp[\mp(-bx/\varepsilon)^{1/2}]$. Putting it roughly, let us say that the second-order equation degenerates to the reduced equation, $by = 0$, (as ε and μ tend to zero) where, in Case (i), the degeneration is through the intermediate equation $\mu a y' + by = 0$; and, in Case (iii), the degeneration is direct (almost as if $\mu = 0$). In order to analyze the limiting behavior of y_1 and y_2, then, it is essential to know the limit of ε/μ^2; i.e., ε and μ must be interrelated in order to ascertain the limiting behavior of the asymptotic solutions.

In this section, we shall only consider the first (but, perhaps, the most interesting) case appropriate for the mth-order equation (3.45), namely

$$\frac{\varepsilon}{\mu^{(m-p)/(n-p)}} \to 0 \qquad \text{as} \quad \mu \to 0. \tag{3.48}$$

Here it is convenient to introduce the new small positive parameters

$$\zeta = \mu^{1/(n-p)} \tag{3.49a}$$

and

$$\eta = \left[\frac{\varepsilon}{\mu^{(m-p)/(n-p)}}\right]^{1/(m-n)} \tag{3.49b}$$

in terms of which we have

$$\varepsilon = \zeta^{m-p}\eta^{m-n} \qquad \text{and} \qquad \mu = \zeta^{n-p}.$$

We shall now proceed treating ζ and η as independent small parameters. Note first that the order of the differential equation (3.45) drops from m when ζ and η are both positive to $p < m$ when $\zeta = \eta = 0$. From our successful analysis of one-parameter problems, we might expect (under appropriate hypotheses) that the limiting solution within (0, 1) as ζ and η both tend to zero will be the solution of a reduced problem consisting of the reduced equation [(3.45) with $\varepsilon = \mu = 0$] and p of the original m boundary conditions (3.46). A cancellation law will be needed to determine which $m - p$ boundary conditions are omitted in defining this reduced problem.

We shall assume that the coefficients in the differential equation (3.45) are infinitely differentiable throughout $0 \leq x \leq 1$ and that

$$\beta(x) \neq 0 \neq \gamma(x) \tag{3.50}$$

there. Then we might seek m linearly independent asymptotic solutions $Y_j(x, \zeta, \eta)$ of the form

$$Y_j(x, \zeta, \eta) = G_j(x, \zeta, \eta)\exp\left[\frac{1}{\zeta\eta}\int_{x_j}^{x} v_j(t, \zeta, \eta)\, dt\right]$$

[generalizing the previous representation (3.22)]. Somewhat more efficiently, however, we shall find $m - n$ asymptotic solutions of the form

$$G_j(x, \zeta\eta, \eta^{n-p})\exp\left[\frac{1}{\zeta\eta}\int_{x_j}^{x} v_j(s, \eta^{n-p})\, ds\right], \qquad j = 1, 2, \ldots, m - n, \tag{3.51}$$

$n - p$ asymptotic solutions of the form

$$G_j(x, \zeta, \eta^{m-n})\exp\left[\frac{1}{\zeta}\int_{x_j}^{x} v_j(s, \eta^{m-n})\, ds\right], \qquad j = m - n + 1, \ldots, m - p, \tag{3.52}$$

and p asymptotic solutions of the form

$$G_j(x, \zeta^{n-p}, \zeta^{m-p}\eta^{m-n}), \qquad j = m - p + 1, \ldots, m. \tag{3.53}$$

Here the $G_j(x, \alpha, \beta)$'s are to have double asymptotic expansions

$$G_j(x, \alpha, \beta) \sim \sum_{r,s=0}^{\infty} G_{jrs}(x)\, \alpha^r \beta^s;$$

i.e., for every integer $N \geq 0$,

$$G_j(x, \alpha, \beta) = \sum_{\substack{r>0, s>0 \\ r+s \leq N}} G_{jrs}(x)\, \alpha^r \beta^s + o[(\max(\alpha, \beta))^N].$$

Further, the exponents have asymptotic expansions in the indicated parameters η^{n-p} or η^{m-n}. Formally substituting (3.51), then, into the differential equation (3.45), we have

$$\frac{1}{\zeta^p \eta^n}\Bigg[\bigg(G_j v_j^m + \zeta\eta\bigg(mG'_j v_j^{m-1} + \frac{m(m-1)}{2} G_j v_j^{m-2} v'_j\bigg)$$

$$+ (\zeta\eta)^2(\cdots) + (\zeta\eta)^m G_j^{(m)}\bigg) + \alpha_1(x)\zeta\eta G_j v_j^{m-n} + (\zeta\eta)^2(\cdots)\Bigg]$$

$$+ \frac{\beta(x)}{\zeta^p \eta^n}\Bigg[\bigg(G_j v_j^n + \zeta\eta\bigg(nG'_j v_j^{n-1} + \frac{n(n-1)}{2} G_j v_j^{n-2} v'_j\bigg)$$

$$+ (\zeta\eta)^2(\cdots)\bigg) + \beta_1(x)\,\zeta\eta G_j v_j^{n-1} + (\zeta\eta)^2(\cdots)\Bigg] + \frac{\gamma(x)}{\zeta^p \eta^p}\Bigg[\bigg(G_j v_j^p$$

$$+ \zeta\eta\bigg(pG'_j v_j^{p-1} + \frac{p(p-1)}{2} G_j v_j^{p-2} v'_j\bigg) + (\zeta\eta)^2(\cdots) + (\zeta\eta)^p G_j^{(p)}\bigg)$$

$$+ \zeta\eta(\cdots) + (\zeta\eta)^p \gamma_p(x) G_j\Bigg]$$

$$= 0.$$

To eliminate the most singular terms, we ask that v_j satisfy

$$v_j^m + \beta(x)\, v_j^n + \eta^{n-p}\gamma(x)\, v_j^p = 0, \tag{3.54}$$

obtaining $m - n$ distinct roots $v_j(x, \eta^{n-p})$ which have series expansions in η^{n-p} with distinct selections

$$v_j(x,0) = \left(-\beta(x)\right)^{1/m-n}, \qquad j = 1, 2, \ldots, m - n. \tag{3.55}$$

The corresponding G_j's are then obtained by a regular perturbation procedure. In particular, equating the next-order terms in the differential equation to zero, we find that G_{j00} must satisfy the nonsingular linear equation

$$G'_{j00}\left(mv_j^{m-1} + n\,\beta(x)\,v_j^{n-1}\right) + G_{j00}\left(\frac{m(m-1)}{2} v_j^{m-2} v'_j + \frac{n(n-1)}{2}\beta(x)\,v_j^{n-2} v'_j + \alpha_1(x)\,v_j^{m-1} + \beta(x)\,\beta_1(x)\,v_j^{n-1}\right) = 0.$$

Thus we formally obtain $m - n$ asymptotic solutions (3.51). Analogously, for the $n - p$ solutions of the form (3.52), the v_j's must be distinct nonzero roots of the polynomial

$$\beta(x)v_j^n + \gamma(x)v_j^p + \eta^{m-n}v_j^m = 0. \tag{3.56}$$

Thus

$$v_j(x,0) = \left(-\frac{\gamma(x)}{\beta(x)}\right)^{1/(n-p)}, \qquad j = m - n + 1, \ldots, m - p \tag{3.57}$$

and the corresponding G_j's can be obtained termwise as solutions of nonsingular linear equations (by a regular perturbation procedure). Finally, the remaining p G_j's of (3.53) are regular perturbations of the p linearly independent solutions of the reduced equation

$$y^{(p)} + \gamma_1(x)y^{(p-1)} + \cdots + \gamma_p(x)y = 0.$$

The m formally obtained solutions so constructed can be shown to be asymptotically valid throughout $0 \le x \le 1$ as ζ and $\eta \to 0$ by generalizing the procedure of Turrittin (1936).

We note that the exponents in the first $m - n$ asymptotic solutions (3.51) approximate the exponents which would result from the simpler equation $\varepsilon y^{(m)} + \mu \beta(x) y^{(n)} = 0$, while the exponents of the $n - p$ asymptotic solutions (3.52) approximate those of $\mu \beta(x) y^{(n)} + \gamma(x) y^{(p)} = 0$. Further, the former exponents are more singular (as ζ and η tend to zero) than the latter since Case (i) applies. In Case (i), then, we say that the mth-order equation (3.45) degenerates to an intermediate nth-order equation and, then, to the reduced pth order equation as ε and μ tend to zero. Neither the intermediate nor the reduced problem will be singular since $\beta(x) \neq 0 \neq \gamma(x)$ throughout $0 \leq x \leq 1$.

We shall call the differential equation (3.45) nonexceptional when all $m - p$ determinations $v_j(x, 0)$ of (3.55) and (3.57) have nonzero real parts. Otherwise, (3.45) is exceptional. Since the limiting behavior is more difficult to analyze in the latter case, we shall not consider it here [see O'Malley (1967b), however]. [We note that Eq. (3.45) will be nonexceptional if both $m - n$ and $n - p$ are odd.] Let us, then, order the roots $v_j(x, 0)$ so that their real parts satisfy

$$\operatorname{Re} v_j(x,0) < 0 \qquad \text{for} \quad j = 1, 2, \ldots, \sigma_1 \tag{3.58a}$$

$$\operatorname{Re} v_j(x,0) > 0 \qquad \text{for } j = \sigma_1 + 1, \ldots, \sigma_1 + \tau_1 = m - n \tag{3.58b}$$

$$\operatorname{Re} v_j(x,0) < 0 \qquad \text{for} \quad j = m - n + 1, \ldots, m - n + \sigma_2 \tag{3.58c}$$

and

$$\operatorname{Re} v_j(x,0) > 0$$
$$\text{for} \quad j = m - n + \sigma_2 + 1, \ldots, m - n + \sigma_2 + \tau_2 = m - p. \tag{3.58d}$$

Thus, we state

THE CANCELLATION LAW: Cancel $\sigma = \sigma_1 + \sigma_2$ boundary conditions at $x = 0$ and $\tau = \tau_1 + \tau_2$ boundary conditions at $x = 1$, starting from those involving the highest derivatives.

Note that the cancellation law is well defined if and only if

$$\sigma \leq r \qquad \text{and} \qquad \tau \leq m - r. \tag{3.59}$$

Slightly modifying Theorem 2 (in the nonexceptional case), then, we have:

COROLLARY: *Suppose*

(i) *the reduced problem*

$$z^{(p)} + \gamma_1(x)z^{(p-1)} + \cdots + \gamma_p(x)z = 0, \qquad 0 \le x \le 1$$
$$z^{(\lambda_i)}(0) = l_i, \qquad i = \sigma + 1, \ldots, r$$
$$z^{(\lambda_i)}(1) = l_i, \qquad i = r + \tau + 1, \ldots, m$$

has a unique solution $z(x)$ *and*

(ii) (a) $\lambda_1, \lambda_2, \ldots, \lambda_{\sigma_1}$ *are distinct modulo* $m - n$,
(b) $\lambda_{\sigma_1+1}, \lambda_{\sigma_1+2}, \ldots, \lambda_\sigma$ *are distinct modulo* $n - p$,
(c) $\lambda_{r+1}, \lambda_{r+2}, \ldots, \lambda_{r+\tau_1}$ *are distinct modulo* $m - n$, *and*
(d) $\lambda_{r+\tau_1+1}, \lambda_{r+\tau_1+2}, \ldots, \lambda_{r+\tau}$ *are distinct modulo* $n - p$.

Then, in the nonexceptional case [where (3.58) holds], the boundary value problem (3.45)–(3.46) has a unique solution $y(x, \varepsilon, \mu)$ *for* ε *and* μ *sufficiently small and with* $\varepsilon/(\mu^{(m-p)/(n-p)}) \to 0$ *as* $\mu \to 0$. *It has the form*

$$\begin{aligned} y(x,\varepsilon,\mu) = {} & \tilde{G}(x,\zeta,\eta) + \zeta^{\lambda_\sigma}\eta^{\lambda_{\sigma_2}} \sum_{j=1}^{\sigma_1} \tilde{G}_j(x,\zeta,\eta) \\ & \times \exp\left[\frac{1}{\zeta\eta}\int_0^x v_j(s,\eta^{n-p})\,ds\right] \\ & + \zeta^{\lambda_{r+\tau}}\eta^{\lambda_{r+\tau_2}} \sum_{j=1}^{\tau_1} \tilde{G}_{\sigma_1+j}(x,\zeta,\eta)\exp\left[\frac{1}{\zeta\eta}\int_1^x v_{\sigma_1+j}(s,\eta^{n-p})\,ds\right] \\ & + \zeta^{\lambda_\sigma} \sum_{j=1}^{\sigma_2} \tilde{G}_{m-n+j}(x,\zeta,\eta) \\ & \times \exp\left[\frac{1}{\zeta}\int_0^x v_{m-n+j}(s,\eta^{m-n})\,ds\right] \\ & + \zeta^{\lambda_{r+\tau}} \sum_{j=1}^{\tau_2} \tilde{G}_{m-n+\sigma_2+j}(x,\zeta,\eta) \\ & \times \exp\left[\frac{1}{\zeta}\int_1^x v_{m-n+\sigma_2+j}(s,\eta^{m-n})\,ds\right], \end{aligned} \tag{3.60}$$

where the functions $\tilde{G}$ and $\tilde{G}_k$ have double asymptotic series expansions in ζ and η throughout $0 \leq x \leq 1$ which may be obtained successively by a scheme of undetermined coefficients. Further $\tilde{G}(x, 0, 0) = z(x)$ and

$$\frac{d^j y(x, \varepsilon, \mu)}{dx^j} \to z^{(j)}(x), \qquad j = 0, 1, 2, \ldots$$

within $0 < x < 1$ as ε and μ tend to zero.

Remarks

1. Note that the expression (3.60) for y can be differentiated any number of times. Also note that since the exponents in (3.60) all tend to $-\infty$ as ε and μ tend to zero, only the outer solution $\tilde{G}(x, \zeta, \eta)$ is asymptotically significant within (0, 1). Moreover, $\tilde{G}$ is a linear combination of the p asymptotic solutions (3.53). Since $\zeta\eta \ll \zeta$, we also note that the first sums in (3.60) decay much faster than the last; i.e., the $n - p$ asymptotic solutions (3.52) are asymptotically more significant than the $m - n$ asymptotic solutions (3.51). [Of course, all m asymptotic solutions are necessary to obtain the uniformly valid representation (3.60).] Finally, we observe that if a specific relationship between ζ and η becomes known (e.g., $\zeta = \eta^3$), the expansion (3.60) can be considerably simplified.

2. As for the theorem, a list of examples can be given for which a hypothesis of the corollary is violated and no limiting solution exists as ε and μ tend to zero.

3. Generalizations of these results to systems of differential equations involving many parameters have also been made [cf. Wasow (1964), Harris (1965), and O'Malley (1969b)].

4. Many-parameter problems can be expected to occur often in practical problems. Among many applications of two-parameter problems, the reader might refer to DiPrima (1968) for one in lubrication theory or to Vasil'eva (1963b) or Chen and O'Malley (1973) for ones in chemical reactor theory and dc motor analysis.

EXAMPLE: Finally, consider the initial value problem

$$\varepsilon y'' + \mu a y' + b y = 0$$
$$y(0) = 1, \qquad y'(0) = 0,$$

where a and b are constants, $x \geq 0$, and the small parameters ε and μ are such that μ and ε/μ^2 both tend to zero. Its unique solution

$$y(x, \varepsilon, \mu)$$
$$= \frac{1}{2(1 - 4\varepsilon b/\mu^2 a^2)^{1/2}}$$
$$\times \left\{\left(-1 + \left(1 - \frac{4\varepsilon b}{\mu^2 a^2}\right)^{1/2}\right) \exp\left[-\frac{\mu a x}{2\varepsilon}\left(1 + \left(1 - \frac{4\varepsilon b}{\mu^2 a^2}\right)^{1/2}\right)\right]\right.$$
$$\left. + \left(1 + \left(1 - \frac{4\varepsilon b}{\mu^2 a^2}\right)^{1/2}\right) \exp\left[-\frac{\mu a x}{2\varepsilon}\left(1 - \left(1 - \frac{4\varepsilon b}{\mu^2 a^2}\right)^{1/2}\right)\right]\right\}$$

tends to the trivial solution of the reduced equation, $by = 0$, in the limit provided both a and b are positive.

As for the general problem (3.45)–(3.46), we could introduce $\zeta = \mu$ and $\eta = \varepsilon/\mu^2$ and then construct asymptotic solutions of the form

$$y_1 = G_1\left(x, \frac{\varepsilon}{\mu}, \frac{\varepsilon}{\mu^2}\right) \exp\left[\frac{\mu}{\varepsilon} \int_0^x v_1\left(s, \frac{\varepsilon}{\mu^2}\right) ds\right]$$

with $v_1(x, 0) = -b$ and

$$y_2 = G_2\left(x, \mu, \frac{\varepsilon}{\mu^2}\right) \exp\left[\frac{1}{\mu} \int_0^x v_2\left(s, \frac{\varepsilon}{\mu^2}\right) ds\right]$$

with $v_2(x, 0) = -b/a$. Both solutions will decay exponentially away from $x = 0$ provided a and b are both positive and we will have $\sigma_1 = \sigma_2 = 1$, $\tau_1 = \tau_2 = 0$. (See Fig. 11.) Note that the "boundary layer thickness" of y_1 is much less then that of y_2. Thus, the corollary

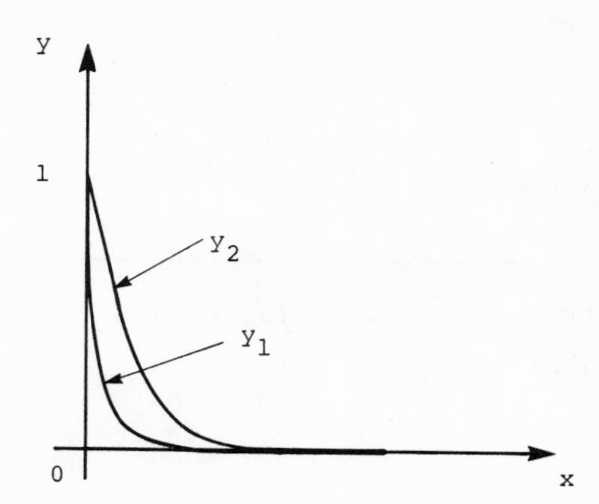

FIGURE 11 The linearly independent solutions y_1 and y_2 of $\varepsilon y'' + \mu a y' + by = 0$, $a > 0$, $b > 0$, $\varepsilon/\mu^2 \to 0$.

implies that the unique solution y will have the form

$$y(x, \varepsilon, \mu) = \frac{\varepsilon}{\mu^2} \tilde{G}_1\left(x, \mu, \frac{\varepsilon}{\mu^2}\right) \exp\left[\frac{\mu}{\varepsilon} \int_0^x v_1\left(s, \frac{\varepsilon}{\mu^2}\right) ds\right]$$
$$+ \tilde{G}_2\left(x, \mu, \frac{\varepsilon}{\mu^2}\right) \exp\left[\frac{1}{\mu} \int_0^x v_2\left(s, \frac{\varepsilon}{\mu^2}\right) ds\right],$$

where the terms in the expansions for $\tilde{G}_1$ and $\tilde{G}_2$ can be obtained by an undetermined coefficients scheme. Note that the outer solution $\tilde{G}(x, \mu, \varepsilon/\mu^2)$ of (3.60) does not appear here because the two linearly independent solutions of this second-order differential equation are accounted for by y_1 and y_2; i.e., $\tilde{G}$ is asymptotically zero. Note, too, that this representation agrees with the known exact solution. This and other boundary value problems for variable coefficient second-order equations are discussed by O'Malley (1967a).

CHAPTER 4

NONLINEAR INITIAL VALUE PROBLEMS

1. THE BASIC PROBLEM

We will consider the nonlinear system

$$
\begin{aligned}
\frac{dx}{dt} &= u(x, y, t, \varepsilon) \\
\varepsilon \frac{dy}{dt} &= v(x, y, t, \varepsilon)
\end{aligned}
\tag{4.1}
$$

of scalar equations for $t \geq 0$ subject to the initial conditions

$$
\begin{aligned}
x(0, \varepsilon) &= x^0(\varepsilon) \\
y(0, \varepsilon) &= y^0(\varepsilon).
\end{aligned}
\tag{4.2}
$$

Here ε is a small positive parameter, u, v, x^0, and y^0 have the asymptotic expansions

$$u(x, y, t, \varepsilon) \sim \sum_{j=0}^{\infty} u_j(x, y, t)\varepsilon^j$$

$$v(x, y, t, \varepsilon) \sim \sum_{j=0}^{\infty} v_j(x, y, t)\varepsilon^j$$

$$x^0(\varepsilon) \sim \sum_{j=0}^{\infty} x_j^0 \varepsilon^j$$

$$y^0(\varepsilon) \sim \sum_{j=0}^{\infty} y_j^0 \varepsilon^j$$

as $\varepsilon \to 0$, and the u_j's and v_j's are infinitely differentiable functions of x, y, and t.

The reduced problem is

$$\begin{aligned} \frac{dx}{dt} &= u_0(x, y, t) \\ 0 &= v_0(x, y, t), \qquad t \geq 0 \\ x(0) &= x_0^0. \end{aligned} \tag{4.3}$$

We shall make the two assumptions

(Hi) *that there is a continuously differentiable function* $\phi(X, t)$ *such that*

$$v_0\big(X, \phi(X, t), t\big) = 0 \tag{4.4}$$

and that the resulting nonlinear initial value problem

$$\begin{aligned} \frac{dx}{dt} &= u_0\big(x, \phi(x, t), t\big) \equiv \tilde{u}(x, t) \\ x(0) &= x_0^0 \end{aligned} \tag{4.5}$$

has a (unique) solution $X_0(t)$ *on some closed bounded interval, say* $0 \leq t \leq 1$, *such that*

$$v_{0y}\big(X_0(t), Y_0(t), t\big) \leq -\kappa \tag{4.6}$$

on $0 \leq t \leq 1$ *for some constant* $\kappa > 0$ *and for*

$$Y_0(t) \equiv \phi\big(X_0(t), t\big) \tag{4.7}$$

and

(Hii) *that for the same* $\kappa > 0$,

$$v_{0y}\big(X_0(0), \lambda, 0\big) \leq -\kappa \tag{4.8}$$

for all values λ *between* $Y_0(0)$ *and* y_0^0.

Under these hypotheses, we shall be able to construct an asymptotic solution of the initial value problem (4.1)–(4.2) which has $(X_0(t), Y_0(t))$ as its limiting solution for $t > 0$ as $\varepsilon \to 0$. Since $X_0(0) = x_0^0$ and $Y_0(0) \neq y_0^0$, in general, $x(t, \varepsilon)$ will converge uniformly in $[0, 1]$ as $\varepsilon \to 0$ but convergence of $y(t, \varepsilon)$ will usually be nonuniform at $t = 0$.

Note that the pair

$$\big(X_0(t), Y_0(t)\big)$$

satisfies the reduced problem (4.3) and, by (Hi) and the implicit function theorem [see, e.g., Hale (1969)], the reduced problem has no other nearby solution. Note, too, that if the reduced problem (4.3) has two (or more) distinct solutions $(X_0(t), Y_0(t))$ corresponding to different choices for $\phi(X, t)$, condition (4.8) can only hold for one of them. Finally, observe that (4.8) follows by continuity from (4.6) for small values of the "boundary layer jump" $J \equiv y_0^0 - Y_0(0)$. It also follows from (4.6) when u_0 is linear in y. In general, however, (4.6) and (4.8) are independent hypotheses.

To illustrate the hypotheses and to predict the form of solution obtained, we consider the linear problem

$$\varepsilon x'' + a(t)x' + b(t)x = 0, \qquad 0 \leq t \leq 1$$

$$x(0) = \alpha, \qquad x'(0) = \gamma,$$

where a and b are infinitely differentiable and $a(t) > 0$. We put this into the system form as

$$\frac{dx}{dt} = y$$

$$\varepsilon \frac{dy}{dt} = -b(t)x - a(t)y, \qquad 0 \leq t \leq 1$$

$$x(0, \varepsilon) = \alpha, \qquad y(0, \varepsilon) = \gamma.$$

As we showed in Chapter 2, this problem has a unique asymptotic solution

$$x(t, \varepsilon) = \hat{A}(t, \varepsilon) + \varepsilon \hat{B}(t, \varepsilon) \exp\left[-\frac{1}{\varepsilon}\int_0^t a(s)\, ds\right]$$

$$y(t, \varepsilon) = \hat{A}'(t, \varepsilon) + [-a(t)\hat{B}(t, \varepsilon) + \varepsilon \hat{B}'(t, \varepsilon)] \times \exp\left[-\frac{1}{\varepsilon}\int_0^t a(s)\, ds\right]$$

where $\hat{A}$ and $\hat{B}$ have asymptotic series expansions as $\varepsilon \to 0$ with infinitely differentiable coefficients and

$$\hat{A}(t, 0) = \alpha \exp\left[-\int_0^x \frac{b(s)}{a(s)}\, ds\right].$$

Introducing the stretched variable

$$\tau = t/\varepsilon,$$

note that $e^{-a(0)\tau}$ decays exponentially to zero as $\varepsilon \to 0$ for any fixed $t > 0$. Further,

$$\exp\left[-\frac{1}{\varepsilon}\int_0^t \Big(a(s) - a(0)\Big)\, ds\right]$$

can be expanded as a power series in ε with coefficients which are polynomials in τ. Expanding the relevant functions of $t = \varepsilon\tau$ as functions of τ, we can rewrite the solution of the initial value problem in the form

$$x(t,\varepsilon) = \hat{A}(t,\varepsilon) + \varepsilon p_1(\tau,\varepsilon)\, e^{-a(0)\tau}$$
$$y(t,\varepsilon) = \hat{A}'(t,\varepsilon) + p_2(\tau,\varepsilon)\, e^{-a(0)\tau},$$

where

$$p_i(\tau,\varepsilon) \sim \sum_{j=0}^{\infty} p_{ij}(\tau)\varepsilon^j,\ i = 1, 2, \qquad \text{as} \quad \varepsilon \to 0$$

and each p_{ij} is a polynomial in τ. Note that, when $t \geq \delta > 0$, $\tau \to \infty$ and

$$p_i(\tau,\varepsilon)e^{-a(0)\tau} = O(\varepsilon^M) \qquad \text{as} \quad \varepsilon \to 0$$

for any M. Away from $t = 0$, then, (x,y) will converge to the outer solution $(\hat{A},\hat{A}')$ as $\varepsilon \to 0$. Moreover, the solution is expressed as the sum of an outer solution (a function of t having an asymptotic series expansion as $\varepsilon \to 0$) plus a boundary layer correction (a function of τ having an asymptotic expansion as $\varepsilon \to 0$ with terms tending to zero as $\tau \to \infty$). We note that the reduced problem for this linear system

$$\frac{dx}{dt} = y, \qquad x(0) = \alpha$$
$$0 = -b(t)x - a(t)y.$$

Further, for this system, ϕ is uniquely given by

$$\phi(x,t) = -\frac{b(t)}{a(t)}x$$

while

$$v_{0y}(x,y,t) = -a(t) < 0$$

for $0 \leq t \leq 1$ and the solution of the reduced problem is given by

$$\left(\hat{A}(t,0), \hat{A}'(t,0)\right) = \left(\alpha \exp\left[-\int_0^x \frac{b(s)}{a(s)} ds\right], \right.$$
$$\left. -\alpha \frac{b(t)}{a(t)} \exp\left[-\int_0^x \frac{b(s)}{a(s)} ds\right]\right).$$

As in the linear example, the asymptotic solution of (4.1)–(4.2) will be an additive function of the variable t and of the stretched variable

$$\tau = t/\varepsilon, \tag{4.9}$$

which tends to infinity as $\varepsilon \to 0$ whenever $t \geq \delta > 0$, δ arbitrary but fixed. Use of this stretched variable will allow us to describe the nonuniform convergence at $t = 0$. We shall seek a solution of the form

$$\begin{aligned} x(t,\varepsilon) &= X(t,\varepsilon) + \varepsilon m(\tau,\varepsilon) \\ y(t,\varepsilon) &= Y(t,\varepsilon) + n(\tau,\varepsilon), \end{aligned} \tag{4.10}$$

where X, Y, m, and n all have asymptotic series expansions as $\varepsilon \to 0$; i.e.,

$$X(t,\varepsilon) \sim \sum_{j=0}^{\infty} X_j(t)\varepsilon^j, \qquad Y(t,\varepsilon) \sim \sum_{j=0}^{\infty} Y_j(t)\varepsilon^j$$
$$m(\tau,\varepsilon) \sim \sum_{j=0}^{\infty} m_j(\tau)\varepsilon^j, \qquad n(\tau,\varepsilon) \sim \sum_{j=0}^{\infty} n_j(\tau)\varepsilon^j.$$

We shall ask that all terms in the expansions for m and n tend to zero as τ tends to infinity. (This condition replaces the familiar matching (patching) conditions used elsewhere.) Away from $t = 0$, then, (x,y) will converge to the outer solution (X, Y) as $\varepsilon \to 0$, while the boundary layer correction $(\varepsilon m, n)$ will be significant only near $t = 0$. Our expansion procedure will consist of obtaining, first, the outer expansion (i.e., the expansion of the outer solution) and, second, the complete expansion.

Since the solution is asymptotically given by the outer solution for $t > 0$, $(X(t,\varepsilon), Y(t,\varepsilon))$ must formally satisfy the system (4.1). Substituting into (4.1), we have equality when $\varepsilon = 0$ since (X_0, Y_0) is a solution of the reduced problem. Using Taylor series expansions of u and v about $(X_0, Y_0, t, 0)$ and equating coefficients of ε^j for $j > 0$, we must have

$$\frac{dX_j}{dt} = u_{0x}(X_0, Y_0, t)X_j + u_{0y}(X_0, Y_0, t)Y_j + \alpha_{j-1}(t)$$
$$0 = v_{0x}(X_0, Y_0, t)X_j + v_{0y}(X_0, Y_0, t)Y_j + \beta_{j-1}(t),$$

where $\alpha_0(t) = u_1(X_0, Y_0, t)$, $\beta_0(t) = v_1(X_0, Y_0, t) - dY_0/dt$, and, generally, α_{j-1} and β_{j-1} are known successively in terms of the X_l's and Y_l's with $l < j$. By assumption (Hi),

$$\phi_x(X_0, t) = \frac{-v_{0x}(X_0, Y_0, t)}{v_{0y}(X_0, Y_0, t)}$$

is defined throughout $0 \leq t \leq 1$, so we have

$$\begin{aligned} Y_j(t) &= \phi_x\big(X_0(t), t\big)X_j(t) + \tilde{\beta}_{j-1}(t) \\ \frac{dX_j(t)}{dt} &= \tilde{u}_x\big(X_0(t), t\big)X_j(t) + \tilde{\alpha}_{j-1}(t), \end{aligned} \tag{4.11}$$

where $\tilde{\alpha}_{j-1}$ and $\tilde{\beta}_{j-1}$ are known successively and $\tilde{u}$ is defined by (4.5). We will obtain X_j (and thereby Y_j) uniquely as the solution of its linear equation once the initial value

$$X_j(0) = x_j^0 - m_{j-1}(0)$$

is determined. Thus, the outer expansion (X, Y) can be completely determined termwise provided the constant $m_{j-1}(0)$ is known at each step. We wish to emphasize that, to obtain the outer solution, it is not sufficient to know the prescribed initial value $x^0(\varepsilon)$. One also needs the initial value $m(0,\varepsilon)$ of the boundary layer correction.

The complete expansion (4.10) must also satisfy the system (4.1). Substituting (4.10) into (4.1), we find that the boundary layer correction $(\varepsilon m, n)$ must satisfy the nonlinear system

$$
\begin{aligned}
\frac{dm}{d\tau} &= \frac{dx}{dt} - \frac{dX}{dt} \\
&= u\big(X(\varepsilon\tau,\varepsilon) + \varepsilon m(\tau,\varepsilon), Y(\varepsilon\tau,\varepsilon) + n(\tau,\varepsilon), \varepsilon\tau, \varepsilon\big) \\
&\quad - u\big(X(\varepsilon\tau,\varepsilon), Y(\varepsilon\tau,\varepsilon), \varepsilon\tau, \varepsilon\big) \\
\frac{dn}{d\tau} &= \varepsilon\left(\frac{dy}{dt} - \frac{dY}{dt}\right) \\
&= v\big(X(\varepsilon\tau,\varepsilon) + \varepsilon m(\tau,\varepsilon), Y(\varepsilon\tau,\varepsilon) + n(\tau,\varepsilon), \varepsilon\tau, \varepsilon\big) \\
&\quad - v\big(X(\varepsilon\tau,\varepsilon), Y(\varepsilon\tau,\varepsilon), \varepsilon\tau, \varepsilon\big).
\end{aligned}
\tag{4.12}
$$

Likewise, (4.2) implies the initial condition

$$n(0,\varepsilon) = y^0(\varepsilon) - Y(0,\varepsilon).$$

We shall obtain the expansions for m and n by expanding both sides of (4.12) and equating coefficients of like powers of ε for finite values of τ.

At $\varepsilon = 0$, we have the nonlinear system

$$
\begin{aligned}
\frac{dm_0}{d\tau} &= u_0\big(X_0(0), Y_0(0) + n_0(\tau), 0\big) - u_0\big(X_0(0), Y_0(0), 0\big) \\
&\equiv n_0(\tau)U\big(n_0(\tau)\big) \\
\frac{dn_0}{d\tau} &= v_0\big(X_0(0); Y_0(0) + n_0(\tau), 0\big) - v_0\big(X_0(0), Y_0(0), 0\big) \\
&\equiv n_0(\tau)V\big(n_0(\tau)\big),
\end{aligned}
\tag{4.13}
$$

where U and V, by the mean value theorem, are appropriate partial derivatives. Note that the initial condition implies that $n_0(0) = J \equiv y_0^0 - Y_0(0)$, the boundary layer jump, and $V(J) \le -\kappa$ by assumption (Hii). By the differential equation, then, $|n_0(\tau)|$ is initially decreasing. By assumption (Hii), moreover, $V(n_0(\tau))$ will remain negative and the differential equation implies that $|n_0(\tau)|$ will decrease monotonically to zero as $\tau \to \infty$ such that

$$|n_0(\tau)| \leq |n_0(0)|e^{-\kappa\tau}.$$

Direct solution of the nonlinear differential equation for n_0 is seldom possible, but the integral equation

$$n_0(\tau) = n_0(0) + \int_0^\tau n_0(s)\, V\Big(n_0(s)\Big)\, ds \tag{4.14}$$

on the semi-infinite interval $\tau \geq 0$ can be uniquely solved by successive approximations [cf. Erdélyi (1964)]. (We note that it is essential here that κ be positive.) Knowing n_0, we have

$$m_0(\tau) = -\int_\tau^\infty n_0(s)\, U\Big(n_0(s)\Big)\, ds = O(e^{-\kappa\tau}) \tag{4.15}$$

since $m_0 \to 0$ as $\tau \to \infty$. Then, $X_1(0) = x_1^0 - m_0(0)$ is known, so the terms $X_1(t)$ and $Y_1(t)$ in the outer expansion are uniquely determined.

Expanding the right sides of (4.12) about $(x, y, t, \varepsilon) = (X_0(0), Y_0(0) + n_0(\tau), 0, 0)$, and equating coefficients of ε^j for each $j > 0$, we ask that m_j and n_j satisfy the linear system

$$\begin{aligned} \frac{dm_j}{d\tau} &= u_{0y}\Big(X_0(0), Y_0(0) + n_0(\tau), 0\Big)\, n_j(\tau) + A_{j-1}(\tau) \\ \frac{dn_j}{d\tau} &= v_{0y}\Big(X_0(0), Y_0(0) + n_0(\tau), 0\Big)\, n_j(\tau) + B_{j-1}(\tau), \end{aligned} \tag{4.16}$$

where A_{j-1} and B_{j-1} are known successively and, by induction, satisfy

$$A_{j-1}(\tau) = O\Big(e^{-\kappa(1-\delta)\tau}\Big) = B_{j-1}(\tau) \qquad \text{as} \quad \tau \to \infty$$

for any $\delta > 0$. Since

$$n_j(0) = y_j^0 - Y_j(0)$$

is known successively, we can integrate the linear equation to uniquely obtain $n_j(\tau)$. Then $dm_j/d\tau$ is determined and we have

$$m_j(\tau) = -\int_\tau^\infty \frac{dm_j}{d\tau}(s)\, ds.$$

Moreover, since $u_{0y}(X_0(0), Y_0(0) + n_0(\tau), 0) \leq -\kappa < 0$,

$$n_j(\tau) = O(e^{-\kappa(1-\delta)\tau}) = m_j(\tau) \quad \text{as} \quad \tau \to \infty;$$

i.e., both m_j and n_j are "functions of boundary layer type" in the terminology of Vishik and Lyusternik (1957). Finally, since

$$X_{j+1}(0) = x_{j+1}^0 - m_j(0),$$

the $(j + 2)$nd terms in the outer expansion are now completely specified. Thus, we are able to recursively determine the complete expansion termwise. Summarizing, we have

THEOREM 3: *For each integer $N \geq 0$, the initial value problem* (4.1)–(4.2) *under the hypotheses* (Hi)–(Hii) *has a unique solution for ε sufficiently small which is such that*

$$\begin{aligned} x(t, \varepsilon) &= X_0(t) + \sum_{j=1}^{N} \left(X_j(t) + m_{j-1}(t/\varepsilon)\right)\varepsilon^j + \varepsilon^{N+1} R(t, \varepsilon) \\ y(t, \varepsilon) &= \sum_{j=0}^{N} \left(Y_j(t) + n_j(t/\varepsilon)\right)\varepsilon^j + \varepsilon^{N+1} S(t, \varepsilon), \end{aligned} \tag{4.17}$$

where $R(t, \varepsilon)$ and $S(t, \varepsilon)$ are uniformly bounded throughout $0 \leq t \leq 1$.

Remarks

The expansion technique developed here follows that of O'Malley (197la). With no serious complications, it extends to problems where x and y are vectors. Likewise, by strengthening the hypotheses to achieve asymptotic stability of the solution to the reduced problem, the results are valid on the semi-infinite interval $t \geq 0$ [cf. Hoppensteadt (1966)]. Similarly, Cauchy problems for ordinary differential equations in any Banach space can be studied and used to obtain solutions to initial boundary value problems for partial differential equations [cf. Trenogin (1963, 1970), Hoppensteadt (1970, 1971), and Krein (1971)]. Likewise, Miranker (1973) has used similar methods to develop a numerical scheme for integrating stiff differential equations, and Murphy (1967) gave techniques for boundary layer integrations. As a direct application of Theorem 3, the reader should note the paper by Heinekin *et al.* (1967) which presents a problem in enzyme kinetics and the paper by Hoppensteadt (1974) which discusses an interesting genetics model.

If v_{0y} were positive, the solution of the initial value problem would become unbounded as the positive parameter $\varepsilon \to 0$ for $t > 0$. The corresponding terminal value problem, however, would be well behaved for $\varepsilon \to 0$.

EXAMPLE: We will now consider the nonlinear problem

$$\frac{dx}{dt} = xy$$
$$\varepsilon \frac{dy}{dt} = -y^3 + y$$

with $x(0) = x^0(\varepsilon)$ and $y(0) = y^0(\varepsilon)$ prescribed. Here the reduced problem

$$\frac{dX}{dt} = XY, \qquad -Y^3 + Y = 0$$
$$X(0) = x_0^0$$

has the three solutions:

$$X_0^{(1)}(t) = x_0^0 e^t, \qquad Y_0^{(1)}(t) = 1$$
$$X_0^{(2)}(t) = x_0^0 e^{-t}, \qquad Y_0^{(2)}(t) = -1$$
$$X_0^{(3)}(t) = x_0^0, \qquad Y_0^{(3)}(t) = 0.$$

For $y_0^0 > \sqrt{3}/3$, we can construct a unique asymptotic solution $(x^{(1)}(t,\varepsilon), y^{(1)}(t,\varepsilon))$ of the form (4.17) converging to $(X_0^{(1)}(t), Y_0^{(1)}(t))$ for $0 < t \leq 1$ since

(Hi) $v_{0y}(X_0^{(1)}(t), Y_0^{(1)}(t), t) = -2 < 0$

and

(Hii) $v_{0y}(x_0^0, y, 0) = -3y^2 + 1 < 0$

for y between y_0^0 *and* $Y_0^{(1)}(0) = 1$.

Similarly, for $y_0^0 < -\sqrt{3}/3$, there is a unique asymptotic solution of the form (4.17) converging to $(X_0^{(2)}(t), Y_0^{(2)}(t))$ for $0 < t \leq 1$. Since

$$v_{0y}\left(X_0^{(3)}(t), Y_0^{(3)}(t), t\right) = 1 > 0,$$

we would expect any solution to converge to $(X_0^{(3)}(t), Y_0^{(3)}(t))$ as $\varepsilon \to 0$ unless $y^0(\varepsilon) = 0$.

To be more explicit, we will obtain the first few terms of the asymptotic solution $(x^{(1)}(t,\varepsilon), y^{(1)}(t,\varepsilon))$. For notational simplicity, we will drop the superscripts. The appropriate outer solution will have an expansion

$$X(t,\varepsilon) = x_0^0 e^t + \varepsilon X_1(t) + \varepsilon^2 X_2(t) + \cdots$$
$$Y(t,\varepsilon) = 1 + \varepsilon Y_1(t) + \varepsilon^2 Y_2(t) + \cdots$$

which satisfies the original system for $t > 0$. Equating coefficients of ε^j in the differential equations, for each $j > 0$, we have

$$\frac{dX_1}{dt} = x_0^0 e^t Y_1 + X_1$$
$$0 = -2Y_1,$$

$$\frac{dX_2}{dt} = x_0^0 e^t Y_2 + X_2 + X_1 Y_1$$
$$0 = -2Y_2 - 3Y_1^2 - \frac{dY_1}{dt},$$

etc. Integration then implies

$$Y_1(t) = 0$$
$$X_1(t) = X_1(0)e^t$$

and

$$Y_2(t) = 0$$
$$X_2(t) = X_2(0)e^t$$

with $X_1(0)$ and $X_2(0)$ still undetermined. Unless, by chance, $y_0(\varepsilon) = 1 + O(\varepsilon^3)$, then, the outer solution cannot be a valid representation of the solution near $t = 0$. To account for the nonuniform convergence there, we shall represent the solution in the form

$$x(t, \varepsilon) = X(t, \varepsilon) + \varepsilon m(\tau, \varepsilon)$$
$$y(t, \varepsilon) = Y(t, \varepsilon) + n(\tau, \varepsilon),$$

where τ is the stretched variable $\tau = t/\varepsilon$ and m and n tend to zero as τ tends to infinity. Substituting into the original system [which is also satisfied by the outer solution (X, Y)], we find that the boundary layer corrections m and n must satisfy the nonlinear equations

$$\frac{dm}{d\tau} = X(\varepsilon\tau, \varepsilon)n(\tau, \varepsilon) + \varepsilon\Big(Y(\varepsilon\tau, \varepsilon) + n(\tau, \varepsilon)\Big)\, m(\tau, \varepsilon)$$
$$\frac{dn}{d\tau} = \Big(1 - 3Y^2(\varepsilon\tau, \varepsilon)\Big)\, n(\tau, \varepsilon) - 3Y(\varepsilon\tau, \varepsilon)\, n^2(\tau, \varepsilon) - n^3(\tau, \varepsilon).$$

Likewise, by the initial conditions,

$$x^0(\varepsilon) = X(0, \varepsilon) + \varepsilon m(0, \varepsilon)$$
$$y^0(\varepsilon) = Y(0, \varepsilon) + n(0, \varepsilon).$$

Substituting

$$m(\tau, \varepsilon) = m_0(\tau) + \varepsilon m_1(\tau) + \cdots$$
$$n(\tau, \varepsilon) = n_0(\tau) + \varepsilon n_1(\tau) + \cdots$$

into the system, then, we successively obtain equations for the coefficients (m_j, n_j). In particular, since $Y_0(t) = 1$,

$$\frac{dm_0}{d\tau} = X_0(0)n_0(\tau)$$

$$\frac{dn_0}{d\tau} = -n_0(\tau)\bigl(2 + 3n_0(\tau) + n_0^2(\tau)\bigr)$$

and

$$n_0(0) = y_0^0 - 1$$

$$X_1(0) = x_1^0 - m_0(0).$$

Further, since

$$\left.\frac{dn_0}{d\tau}\right|_{\tau=0} = -n_0(0)(y_0^0 + y_0^{0\,2}),$$

$|n_0(\tau)|$ is initially decreasing provided $y_0^0 > 0$, and, since $2 + 3n + n^2 \geq \kappa + \kappa^2 > \kappa$ for $n > -1 + \kappa > -1$,

$$|n_0(\tau)| \leq |n_0(0)|e^{-\kappa\tau} \qquad \text{for all} \quad \tau \geq 0,$$

provided $y_0^0 \geq \kappa > 0$. This assures us that we can obtain $n_0(\tau)$ by solving the nonlinear integral equation

$$n_0(\tau) = (y_0^0 - 1) - \int_0^\tau n_0(s)\bigl(2 + 3n_0(s) + n_0^2(s)\bigr)\,ds, \qquad \tau \geq 0$$

by successive approximations. [For this particular problem, we can actually integrate the differential equation for y^2 as a Riccati equation from which it follows that

$$n_0(\tau) = \left(1 - \left(\frac{1 - y_0^{0\,2}}{y_0^{0\,2}}\right)e^{-2\tau}\right)^{-1/2} - 1$$

$$= O(e^{-2\tau}) \qquad \text{as} \quad \tau \to \infty.\,]$$

Knowing n_0 and asking that $m_0 \to 0$ as $\tau \to \infty$, we have

$$m_0(\tau) = -x_0^0 \int_\tau^\infty n_0(s)\,ds = O(e^{-2\tau}).$$

In particular, note that this provides the initial value

$$X_1(0) = x_1^0 + x_0^0 \int_0^\infty n_0(s)\,ds$$

needed for the outer expansion.

Continuing, the equations for m_1 and n_1 form the linear system

$$\frac{dm_1}{d\tau} = x_0^0 n_1(\tau) + \left(X_1(0) + \tau x_0^0\right) n_0(\tau) + \left(1 + n_0(\tau)\right) m_0(\tau)$$

$$\frac{dn_1}{d\tau} = -\left(2 + 6n_0(\tau) + 3n_0^2(\tau)\right) n_1(\tau)$$

$$n_1(0) = y_1^0.$$

Integrating, then,

$$n_1(\tau) = y_1^0 e^{-2\tau} \exp\left[-3 \int_0^\tau \left(2n_0(s) + n_0^2(s)\right) ds\right]$$

and, since $m_1 \to 0$ as $\tau \to \infty$,

$$m_1(\tau) = -\int_\tau^\infty [x_0^0 n_1(s) + \left(X_1(0) + s x_0^0\right) n_0(s) + \left(1 + n_0(s)\right) m_0(s)]\,ds$$

is also exponentially decaying as $\tau \to \infty$. Taking $X_2(0) = x_2^0 - m_1(0)$, we completely determine the third-order terms in the outer expansion.

From only partial results, then, we have obtained

$$x(t, \varepsilon) = x_0^0 e^t + \varepsilon\left[\left(x_1^0 + x_0^0 \int_0^\infty n_0(s)\,ds\right)e^t - x_0^0 \int_{t/\varepsilon}^\infty n_0(s)\,ds\right] + O(\varepsilon^2)$$

$$y(t, \varepsilon) = 1 + n_0\left(\frac{t}{\varepsilon}\right) + \varepsilon\left(y_1^0 e^{-2t/\varepsilon} \exp\left[-3 \int_0^{t/\varepsilon} \left(2n_0(s) + n_0^2(s)\right) ds\right]\right) + O(\varepsilon^2)$$

on any bounded interval $0 \leq t \leq T$ provided $y_0^0 > 0$. Here

$$n_0(\tau) = \left(1 - \left(\frac{1 - y_0^{0\,2}}{y_0^{0\,2}}\right)e^{-2\tau}\right)^{-1/2} - 1.$$

Note that Theorem 3 guaranteed the existence of the asymptotic solution only for $y_0^0 > \sqrt{3}/3$, while the constructed solution is valid for $y_0^0 > 0$. Similarly, we could obtain a solution converging to $(x_0^0 e^{-t}, -1)$ for all $t > 0$ provided $y_0^0 < 0$, while the theorem implies

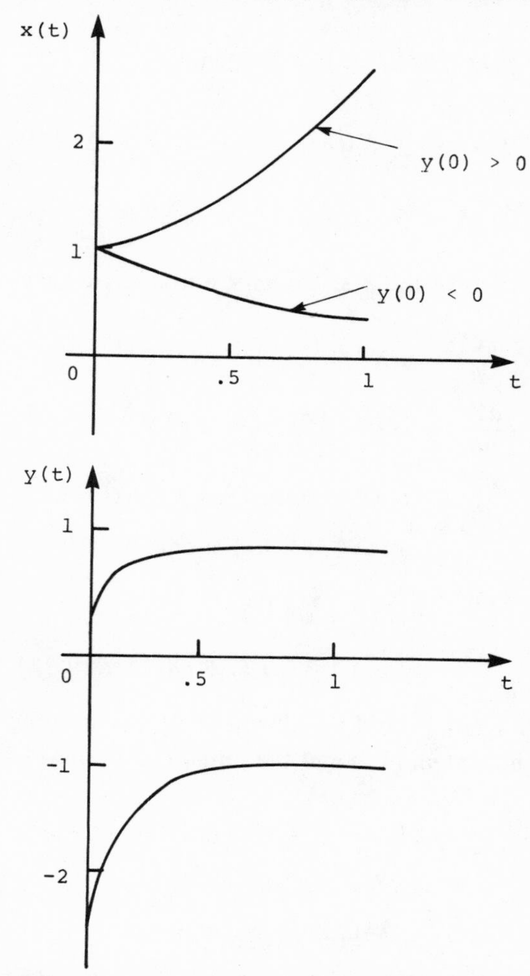

FIGURES 12 AND 13 The solution $x(t)$, $y(t)$ for $dx/dt = xy$, $x(0) = 1$ and $dy/dt = -y^3 + y$, $y(0)$ prescribed in the two cases $y(0) > 0$ and $y(0) < 0$.

this result only on bounded intervals for $y_0^0 < -\sqrt{3}/3$. Thus, it is likely that the expansions of Theorem 3 are valid under assumption (Hi) and a condition somewhat weaker than (Hii). (See Figs. 12 and 13.)

PROOF OF THEOREM 3: Let us define

$$X^N(t,\varepsilon) = \sum_{j=0}^{N} X_j(t)\varepsilon^j, \qquad Y^N(t,\varepsilon) = \sum_{j=0}^{N} Y_j(t)\varepsilon^j$$
$$m^N(\tau,\varepsilon) = \sum_{j=0}^{N-1} m_j(\tau)\varepsilon^j, \qquad n^N(\tau,\varepsilon) = \sum_{j=0}^{N} n_j(\tau)\varepsilon^j.$$

By the definitions of the X_j's, Y_j's, m_j's, and n_j's, we have

$$\begin{aligned}
\frac{dX^N}{dt} &= u(X^N, Y^N, t, \varepsilon) + O(\varepsilon^{N+1}) \\
\varepsilon\frac{dY^N}{dt} &= v(X^N, Y^N, t, \varepsilon) + O(\varepsilon^{N+1}) \\
\frac{dm^N}{d\tau} &= u(X^N + \varepsilon m^N, Y^N + n^N, t, \varepsilon) \\
&\quad - u(X^N, Y^N, t, \varepsilon) + O(\varepsilon^N e^{-\kappa(1-\delta)\tau}) \\
\frac{dn^N}{d\tau} &= v(X^N + \varepsilon m^N, Y^N + n^N, t, \varepsilon) \\
&\quad - v(X^N, Y^N, t, \varepsilon) + O(\varepsilon^{N+1} e^{-\kappa(1-\delta)\tau}),
\end{aligned}$$

where the O symbols hold for all t in [0, 1]. Substituting the solution (4.17) into the system (4.1) and the initial conditions (4.2), we obtain

$$\begin{aligned}
\varepsilon^{N+1}\frac{dR}{dt} &= u(X^N + \varepsilon m^N + \varepsilon^{N+1}R, Y^N + n^N + \varepsilon^{N+1}S, t, \varepsilon) \\
&\quad - u(X^N + \varepsilon m^N, Y^N + n^N, t, \varepsilon) \\
&\quad + O(\varepsilon^{N+1}) + O(\varepsilon^N e^{-\kappa(1-\delta)\tau}) \\
\varepsilon^{N+2}\frac{dS}{dt} &= v(X^N + \varepsilon m^N + \varepsilon^{N+1}R, Y^N + n^N + \varepsilon^{N+1}S, t, \varepsilon) \\
&\quad - v(X^N + \varepsilon m^N, Y^N + n^N, t, \varepsilon) + O(\varepsilon^{N+1}),
\end{aligned}$$

where the O terms are independent of R and S. Further,

$$R(0,\varepsilon) = S(0,\varepsilon) = 0(1).$$

Integrating, then, R and S will satisfy a pair of integral equations of the form

$$R(t,\varepsilon) = R^0(t,\varepsilon) + \int_0^t \tilde{U}\Big(R(\rho,\varepsilon), S(\rho,\varepsilon), \rho, \varepsilon\Big)\, d\rho$$

$$S(t,\varepsilon) = S^0(t,\varepsilon) + \frac{1}{\varepsilon}\int_0^t \tilde{V}\Big(R(\rho,\varepsilon), S(\rho,\varepsilon), \rho, \varepsilon\Big)$$

$$\times \exp\left[\frac{1}{\varepsilon}\int_\rho^t v_y(X^N + \varepsilon m^N, Y^N + n^N, s, \varepsilon)\, ds\right] d\rho,$$

where

$$\tilde{U}(R,S,t,\varepsilon) = \frac{1}{\varepsilon^{N+1}}[u(X^N + \varepsilon m^N + \varepsilon^{N+1}R, Y^N + n^N + \varepsilon^{N+1}S, t, \varepsilon)$$
$$- u(X^N + \varepsilon m^N, Y^N + n^N, t, \varepsilon)],$$

$$\tilde{V}(R,S,t,\varepsilon) = \frac{1}{\varepsilon^{N+1}}[v(X^N + \varepsilon m^N + \varepsilon^{N+1}R, Y^N + n^N + \varepsilon^{N+1}S, t, \varepsilon)$$
$$- v(X^N + \varepsilon m^N, Y^N + n^N, t, \varepsilon)$$
$$- \varepsilon^{N+1} S v_y(X^N + \varepsilon m^N, Y^N + n^N, t, \varepsilon)],$$

$$R^0(t,\varepsilon) = O(1)\left[1 + \int_0^t \left(1 + \frac{1}{\varepsilon}\exp[-\kappa(1-\delta)s/\varepsilon]\right) ds\right] = O(1),$$

and

$$S^0(t,\varepsilon) = O(1) + O\left(\frac{1}{\varepsilon}\right)$$
$$\times \int_0^t \exp\left[\frac{1}{\varepsilon}\int_0^t v_y(X^N + \varepsilon m^N, Y^N + n^N, s, \varepsilon)\, ds\right] dt.$$

By the construction and assumption (Hii), however,

$$v_y(X^N + \varepsilon m^N, Y^N + n^N, t, \varepsilon) \le -\frac{\kappa}{2} < 0$$

provided ε is sufficiently small. Thus, $S^0(t,\varepsilon)$ is bounded throughout $[0, 1]$ and $R(t,\varepsilon)$ and $S(t,\varepsilon)$ can be uniquely determined by successive approximation for ε sufficiently small [cf. Erdélyi (1964) and Willett (1964)]. Thus, there is one solution of (4.1)–(4.2) of the form (4.17).

2. TWO-PARAMETER PROBLEMS

Physical problems frequently involve the solution of boundary value problems with many small parameters. Thus, we are naturally led to consider boundary value problems for nonlinear systems of first-order equations with small parameters multiplying the derivatives.

Before examining initial value problems for such systems, we will first reconsider the linear constant coefficient problem

$$\varepsilon y'' + \mu a y' + b y = 0$$
$$y(0) = 1, \qquad y'(0) = 0$$

on the interval $0 \leq t \leq 1$ where the coefficients a and b are both positive and the small positive parameters ε and μ are such that μ and ε/μ^2 both tend to zero. Recall that we previously showed that the solution converges nonuniformly at $t = 0$ as the parameters tend toward zero and that the limiting solution elsewhere is the trivial solution of the reduced problem.

Introducing the small parameters

$$\varepsilon_1 = \mu \qquad \text{and} \qquad \varepsilon_2 = \varepsilon/\mu^2,$$

we rewrite the second-order equation as

$$\varepsilon_1 \frac{dy}{dt} = z$$
$$\varepsilon_1 \varepsilon_2 \frac{dz}{dt} = -by - az, \qquad 0 \leq t \leq 1,$$

where $y(0) = 1$ and $z(0) = 0$, noting that the reduced system has only the trivial solution $Y_{00}(t) = Z_{00}(t) = 0$. Introducing the stretched variables

$$\tau_1 = \frac{t}{\varepsilon_1} \quad \text{and} \quad \tau_2 = \frac{t}{\varepsilon_1 \varepsilon_2},$$

then, note that both τ_1 and $\tau_2 \to \infty$ for $t > 0$ as ε_1 and $\varepsilon_2 \to 0$, but that τ_2 is then much larger than τ_1. In terms of these variables, we have

$$\begin{aligned} y_{\varepsilon,\mu}(t) = {} & \frac{1}{2(1 - 4\varepsilon_2 b/a^2)^{1/2}} \Big\{ \left(1 + (1 - 4\varepsilon_2 b/a^2)^{1/2}\right) \\ & \times \exp\left[-\frac{a\tau_1}{2\varepsilon_2}\left(1 - \frac{2\varepsilon_2 b}{a^2} - (1 - 4\varepsilon_2 b/a^2)^{1/2}\right)\right] e^{-b\tau_1/a} \\ & + \left(-1 + (1 - 4\varepsilon_2 b/a^2)^{1/2}\right) \\ & \times \exp\left[-\frac{a\tau_2}{2}\left(-1 + (1 - 4\varepsilon_2 b/a^2)^{1/2}\right)\right] e^{-a\tau_2} \Big\}. \end{aligned}$$

More simply, we note that the solution has the form

$$\begin{aligned} y(t; \varepsilon_1, \varepsilon_2) &= \tilde{n}(\tau_1; \varepsilon_2) e^{-b\tau_1/a} + \varepsilon_2 \tilde{g}(\tau_2; \varepsilon_2) e^{-a\tau_2} \\ z(t; \varepsilon_1, \varepsilon_2) &= \tilde{p}(\tau_1; \varepsilon_2) e^{-b\tau_1/a} + \tilde{h}(\tau_2; \varepsilon_2) e^{-a\tau_2}, \end{aligned}$$

where $\tilde{n}(\tau; \varepsilon)$, $\tilde{p}(\tau; \varepsilon)$, $\tilde{g}(\tau; \varepsilon)$, and $\tilde{h}(\tau; \varepsilon)$ all have asymptotic expansions of the form

$$q(\tau; \varepsilon) \sim \sum_{j=0}^{\infty} q_j(\tau)\varepsilon^j, \quad \text{as} \quad \varepsilon \to 0$$

where each q_j is a polynomial in τ of degree $\leq j$. Here, then, the outer expansion (valid away from $t = 0$) is asymptotically zero to all orders $\varepsilon_1^{j_1} \varepsilon_2^{j_2}$. We note that it is essential in this example that the coefficients a and b/a be positive in order to have convergent behavior for $t \geq 0$ as ε_1 and ε_2 both tend to zero.

Now we will seek the asymptotic solution of the more general nonlinear system

$$\begin{aligned} \frac{dx}{dt} &= u(x, y, z, t; \varepsilon_1, \varepsilon_2) \\ \varepsilon_1 \frac{dy}{dt} &= v(x, y, z, t; \varepsilon_1, \varepsilon_2) \\ \varepsilon_1 \varepsilon_2 \frac{dz}{dt} &= w(x, y, z, t; \varepsilon_1, \varepsilon_2) \end{aligned} \tag{4.18}$$

of three scalar equations on the interval $0 \leq t \leq 1$ (or any bounded interval) subject to the initial conditions

$$\begin{aligned} x(0; \varepsilon_1, \varepsilon_2) &= x^0(\varepsilon_1, \varepsilon_2) \\ y(0; \varepsilon_1, \varepsilon_2) &= y^0(\varepsilon_1, \varepsilon_2) \\ z(0; \varepsilon_1, \varepsilon_2) &= z^0(\varepsilon_1, \varepsilon_2) \end{aligned} \tag{4.19}$$

as the small parameters ε_1 and ε_2 both tend to zero. We will assume that u, v, w, x^0, y^0, and z^0 all have double asymptotic expansions in ε_1 and ε_2. For example,

$$u(x, y, z, t; \varepsilon_1, \varepsilon_2) \sim \sum_{j_1, j_2 = 0}^{\infty} u_{j_1 j_2}(x, y, z, t) \varepsilon_1^{j_1} \varepsilon_2^{j_2}.$$

That is, for any integer $J \geq 0$,

$$u(x, y, z, t; \varepsilon_1, \varepsilon_2) = \sum_{\substack{(j_1, j_2) \geq (0,0) \\ j_1 + j_2 \leq J}} u_{j_1 j_2}(x, y, z, t)\, \varepsilon_1^{j_1} \varepsilon_2^{j_2} + O(\lambda^{J+1})$$

$$\text{as} \quad \lambda \equiv \max\{\varepsilon_1, \varepsilon_2\} \to 0.$$

Further, we shall assume that all coefficients in the expansions for u, v, and w are infinitely differentiable.

It might seem somewhat more natural to instead consider systems

$$\begin{aligned} \frac{dx}{dt} &= u(x, y, z, t; \mu, \kappa) \\ \mu \frac{dy}{dt} &= v(x, y, z, t; \mu, \kappa) \\ \kappa \frac{dz}{dt} &= w(x, y, z, t; \mu, \kappa), \end{aligned}$$

where μ and κ are both small parameters. As we shall see, however, the asymptotic solutions can be completely different in the two extremes when $\kappa/\mu \to 0$ as $\mu \to 0$ or $\mu/\kappa \to 0$ as $\kappa \to 0$. In the first case, we have a special system of the form (4.18) with $\varepsilon_1 = \mu$ and $\varepsilon_2 = \kappa/\mu$. In the second, we must interchange the roles of the y and z coordinates in order to obtain a system of the form (4.18). The case when $\mu = c\kappa$ for some nonzero constant c leads to a one-parameter problem of the form (4.1) for a vector y. Analogous considerations apply for the infinite interval problem which describes the "shock layer" limiting behavior of the one-dimensional flow of a fluid when the viscosity and heat conductivity both tend to zero [cf. Gilbarg (1951)].

The reduced problem corresponding to (4.18) is

$$\begin{aligned} \frac{dx}{dt} &= u_{00}(x,y,z,t) \\ 0 &= v_{00}(x,y,z,t) \\ 0 &= w_{00}(x,y,z,t) \\ x(0) &= x_{00}^0. \end{aligned} \tag{4.20}$$

We shall suppose that continuously differentiable functions $\phi(X, Y, t)$ and $\psi(X, t)$ exist such that

$$w_{00}(X, Y, \phi, t) = 0 \tag{4.21}$$

and

$$\tilde{v}(X, \psi, t) \equiv v_{00}\Big(X, \psi, \phi(X, \psi, t), t\Big) = 0 \tag{4.22}$$

and that the (unique) solution $X_{00}(t)$ of the initial value problem

$$\begin{aligned} \frac{dX}{dt} &= \tilde{u}(X,t) \\ X(0) &= x_{00}^0 \end{aligned} \tag{4.23}$$

exists for $0 \leq t \leq 1$, where

$$\tilde{u}(X,t) = u_{00}\Big(X, \psi(X,t), \phi(X, \psi(X,t), t), t\Big).$$

Then

$$\left(X_{00}(t), Y_{00}(t), Z_{00}(t)\right)$$

will be a solution of the reduced problem (4.20) where we define

$$Y_{00}(t) = \psi\left(X_{00}(t), t\right) \tag{4.24}$$

and

$$Z_{00}(t) = \phi\left(X_{00}(t), Y_{00}(t), t\right). \tag{4.25}$$

In order to prove that the original problem (4.18)–(4.19) has a solution converging to (X_{00}, Y_{00}, Z_{00}) away from $t = 0$ as ε_1, $\varepsilon_2 \to 0$, we shall also assume that for some $\kappa > 0$,

$$\begin{aligned} w_{00z}\left(X_{00}(t), Y_{00}(t), Z_{00}(t), t\right) &\leq -\kappa \\ \tilde{v}_y\left(X_{00}(t), Y_{00}(t), t\right) &\leq -\kappa \end{aligned} \tag{4.26}$$

throughout $0 \leq t \leq 1$ and that

$$\begin{aligned} w_{00z}(x_{00}^0, y_{00}^0, \mu, 0) &\leq -\kappa \\ \tilde{v}_y(x_{00}^0, \lambda, 0) &\leq -\kappa \end{aligned} \tag{4.27}$$

for all μ between $\phi(x_{00}^0, y_{00}^0, 0)$ and z_{00}^0 and for all λ between $Y_{00}(0)$ and y_{00}^0. Note that these assumptions are directly analogous to hypotheses (Hi) and (Hii) used for the one-parameter problem, except that $Z_{00}(0) \neq \phi(x_{00}^0, y_{00}^0, 0)$ in general.

Under these conditions, we shall obtain an asymptotic solution of (4.18)–(4.19) of the form

$$\begin{aligned} x(t; \varepsilon_1, \varepsilon_2) &= X(t; \varepsilon_1, \varepsilon_2) + \varepsilon_1 m(\tau_1; \varepsilon_1, \varepsilon_2) + \varepsilon_1 \varepsilon_2 f(\tau_2; \varepsilon_1, \varepsilon_2) \\ y(t; \varepsilon_1, \varepsilon_2) &= Y(t; \varepsilon_1, \varepsilon_2) + n(\tau_1; \varepsilon_1, \varepsilon_2) + \varepsilon_2 g(\tau_2; \varepsilon_1, \varepsilon_2) \\ z(t; \varepsilon_1, \varepsilon_2) &= Z(t; \varepsilon_1, \varepsilon_2) + p(\tau_1; \varepsilon_1, \varepsilon_2) + h(\tau_2; \varepsilon_1, \varepsilon_2), \end{aligned} \tag{4.28}$$

where τ_1 and τ_2 are the stretched variables

$$\tau_1 = t/\varepsilon_1 \qquad \text{and} \qquad \tau_2 = t/\varepsilon_1 \varepsilon_2 \tag{4.29}$$

and X, Y, Z, m, n, p, f, g, and h all have double series expansions in ε_1 and ε_2. Further, m, n, and p tend to zero as $\tau_1 \to \infty$ while f, g, and h tend to zero as $\tau_2 \to \infty$.

The formal expansion can be obtained in three stages by directly generalizing the procedure used for one-parameter problems. First, we determine thc outer expansion

$$\big(X(t; \varepsilon_1, \varepsilon_2), Y(t; \varepsilon_1, \varepsilon_2), Z(t; \varepsilon_1, \varepsilon_2)\big) \tag{4.30}$$

as a formal solution of the system (4.18) with first term $(X_{00}(t), Y_{00}(t), Z_{00}(t))$. Using (4.21), (4.22), and (4.26), higher-order terms will recursively satisfy linear systems with only $X(0; \varepsilon_1, \varepsilon_2)$ unspecified.

Second, we ask that the intermediate expansion

$$\begin{aligned}\big(X(\varepsilon_1 \tau_1; \varepsilon_1, \varepsilon_2) &+ \varepsilon_1 m(\tau_1; \varepsilon_1, \varepsilon_2), Y(\varepsilon_1 \tau_1; \varepsilon_1, \varepsilon_2)\\ &+ n(\tau_1; \varepsilon_1, \varepsilon_2), Z(\varepsilon_1 \tau_1; \varepsilon_1, \varepsilon_2) + p(\tau_1; \varepsilon_1, \varepsilon_2)\big)\end{aligned} \tag{4.31}$$

satisfy the system (4.18) as a function of the lesser stretched variable τ_1. Equating coefficients when $\varepsilon_1 = \varepsilon_2 = 0$, we obtain the nonlinear system

$$\begin{aligned}\frac{dm_{00}}{d\tau_1} &= u_{00}\big(X_{00}(0), Y_{00}(0) + n_{00}(\tau_1), Z_{00}(0) + p_{00}(\tau_1), 0\big)\\ &\quad - u_{00}\big(X_{00}(0), Y_{00}(0), Z_{00}(0), 0\big)\\ \frac{dn_{00}}{d\tau_1} &= v_{00}\big(X_{00}(0), Y_{00}(0) + n_{00}(\tau_1), Z_{00}(0) + p_{00}(\tau_1), 0\big)\\ &\quad - v_{00}\big(X_{00}(0), Y_{00}(0), Z_{00}(0), 0\big)\\ 0 &= w_{00}\big(X_{00}(0), Y_{00}(0) + n_{00}(\tau_1), Z_{00}(0) + p_{00}(\tau_1), 0\big)\\ &\quad - w_{00}\big(X_{00}(0), Y_{00}(0), Z_{00}(0), 0\big)\\ &= w_{00}\big(X_{00}(0), Y_{00}(0) + n_{00}(\tau_1), Z_{00}(0) + p_{00}(\tau_1), 0\big).\end{aligned}$$

Using the definition of ϕ, we take

$$Z_{00}(0) + p_{00}(\tau_1) = \phi\Big(X_{00}(0), Y_{00}(0) + n_{00}(\tau_1), 0\Big)$$

or

$$p_{00}(\tau_1) = \phi\Big(X_{00}(0), Y_{00}(0) + n_{00}(\tau_1), 0\Big) - \phi\Big(X_{00}(0), Y_{00}(0), 0\Big).$$

Then, by (4.22), n_{00} satisfies the nonlinear equation

$$\begin{aligned}\frac{dn_{00}}{d\tau_1} &= \tilde{v}\Big(X_{00}(0), Y_{00}(0) + n_{00}(\tau_1), 0\Big) - \tilde{v}\Big(X_{00}(0), Y_{00}(0), 0\Big)\\ &\equiv n_{00}(\tau_1)\tilde{V}\Big(n_{00}(\tau_1)\Big)\end{aligned}$$

and, by (4.28), $n_{00}(0) = y_{00}^0 - Y_{00}(0)$. By hypothesis (4.27), then, a unique n_{00} can be obtained by use of successive approximations in the nonlinear integral equation

$$n_{00}(\tau_1) = n_{00}(0) + \int_0^{\tau_1} n_{00}(s)\tilde{V}\Big(n_{00}(s)\Big)\, ds, \qquad \tau_1 \geq 0.$$

Further, we will have

$$|n_{00}(\tau_1)| \leq |n_{00}(0)|e^{-\kappa\tau_1} \qquad \text{for} \quad \tau_1 \geq 0.$$

Knowing n_{00}, we also have $p_{00}(\tau_1)$ and

$$m_{00}(\tau_1) = -\int_{\tau_1}^{\infty} \frac{dm_{00}}{d\tau_1}(s)\, ds.$$

Moreover, $p_{00}(\tau_1) = O(e^{-\kappa\tau_1}) = m_{00}(\tau_1)$ as $\tau_1 \to \infty$. Higher-order coefficients m_j, n_j, and p_j will satisfy linear systems and can be completely determined recursively up to specification of $n(0; \varepsilon_1, \varepsilon_2)$. They will also decay exponentially as $\tau_1 \to \infty$.

Third, we ask that the complete expansion

$$\Big(x(\varepsilon_1\varepsilon_2\tau_2; \varepsilon_1, \varepsilon_2), y(\varepsilon_1\varepsilon_2\tau_2; \varepsilon_1, \varepsilon_2), z(\varepsilon_1\varepsilon_2\tau_2; \varepsilon_1, \varepsilon_2)\Big) \tag{4.28}$$

satisfy the system (4.18) as a function of the greater stretched variable τ_2. When $\varepsilon_1 = \varepsilon_2 = 0$, this implies that

$$\frac{df_{00}}{d\tau_2} = u_{00}\Big(x_{00}^0, y_{00}^0, Z_{00}(0) + p_{00}(0) + h_{00}(\tau_2), 0\Big) - u_{00}\Big(x_{00}^0, y_{00}^0, Z_{00}(0) + p_{00}(0), 0\Big)$$

$$\frac{dg_{00}}{d\tau_2} = v_{00}\Big(x_{00}^0, y_{00}^0, Z_{00}(0) + p_{00}(0) + h_{00}(\tau_2), 0\Big) - v_{00}\Big(x_{00}^0, y_{00}^0, Z_{00}(0) + p_{00}(0), 0\Big)$$

$$\frac{dh_{00}}{d\tau_2} = w_{00}\Big(x_{00}^0, y_{00}^0, Z_{00}(0) + p_{00}(0) + h_{00}(\tau_2), 0\Big) - w_{00}\Big(x_{00}^0, y_{00}^0, Z_{00}(0) + p_{00}(0), 0\Big) \equiv h_{00}(\tau_2)\, \tilde{W}\Big(h_{00}(\tau_2)\Big).$$

Note that $h_{00}(0) = z_{00}^0 - Z_{00}(0) - p_{00}(0) = z_{00}^0 - \phi(x_{00}^0, y_{00}^0, 0)$ and the condition (4.27) and the differential equation imply that $|h_{00}(\tau_2)|$ will decay to zero monotonically as τ_2 increases. Moreover, f_{00}, g_{00}, and h_{00} will all be uniquely determined as exponentially decaying terms as $\tau_2 \to \infty$. Higher-order terms of f, g, and h will likewise be recursively determined as exponentially decaying terms up to specification of $h(0; \varepsilon_1, \varepsilon_2)$. Applying the initial conditions for x, y, and z termwise successively yields the unknown initial conditions for X, n, and h. Thus, the expansion (4.28) can be formally obtained. Details of the proof and expansion procedure are given by O'Malley (1971a) (with a reversal in subscript notation).

To summarize, we observe that the asymptotic solution obtained is a sum of functions of t, of $\tau_1 = \tau/\varepsilon_1$, and of $\tau_2 = t/\varepsilon_1\varepsilon_2$. The expansion process first finds the outer expansion depending on t, then constructs the boundary layer correction depending on the least singular stretched variable τ_1, and, finally, adds the contribution due to the most singular stretched variable τ_2. (One often refers to such boundary layer corrections as being of thickness ε_1 and $\varepsilon_1\varepsilon_2$, respectively. The thicker correction is added first.) Many-parameter problems should be treated in analogous fashion. An application of this technique to obtain asymptotic solutions to certain Cauchy problems in a Banach space is given by Gordon (1974), while Chen and O'Malley (1974) treat a problem in chemical flow reactor theory.

3. DIFFERENTIAL–DIFFERENCE EQUATIONS WITH SMALL DELAY

In this section, we wish to consider the initial value problem consisting of the nonlinear differential–difference equation

$$\dot{x}(t) = f\big(t, x(t), x(t-\mu), \dot{x}(t-\mu)\big) \qquad \text{for} \quad t \geq 0 \tag{4.32}$$

and the initial condition

$$x(t) = \phi(t) \qquad \text{for} \quad -\mu \leq t \leq 0 \tag{4.33}$$

as the positive delay parameter μ tends to zero.

We shall assume that

(Hi) *$f(t, x, y, u)$ and $\phi(t)$ are infinitely differentiable in all arguments and that they are independent of μ,*

(Hii) *the nonlinear reduced problem*

$$\begin{aligned} \dot{X}_0(t) &= f\big(t, X_0(t), X_0(t), \dot{X}_0(t)\big) \\ X_0(0) &= \phi(0) \end{aligned} \tag{4.34}$$

has a unique continuously differentiable solution $X_0(t)$ on some interval $0 \leq t \leq T$,

and

(Hiii) *for some $\kappa > 0$,*

$$\big|f_u\big(t, X_0(t), X_0(t), \dot{X}_0(t)\big)\big| < e^{-\kappa} < 1 \tag{4.35a}$$

for $0 \leq t \leq T$, *and*

$$|f_u(0, \phi(0), \phi(0), u)| < e^{-\kappa} < 1 \tag{4.35b}$$

for all u *such that* $|u| \leq |\dot{X}_0(0)| + |\dot{X}_0(0) - \dot{\phi}(0)|$.

[By the implicit function theorem, note that the last assumption implies (among other things) that $\dot{X}_0(t)$ can be uniquely obtained as a function of X_0 and t.] Under these conditions, the solution $x(t, \mu)$ will converge to $X_0(t)$ as $\mu \to 0$ on the closed interval $0 \leq t \leq T$. Derivatives of the solution, however, will converge nonuniformly near $t = 0$.

Before proceeding, we note that the problem could be solved by a stepwise integration scheme for moderate values of μ. We would define

$$x(t) = x_0(t) = \phi(t) \qquad \text{for} \quad -\mu \leq t \leq 0$$

and then, for each integer j with $1 \leq j \leq 1 + T/\mu$, we would set

$$\dot{x}_j(t) = f(t, x_j(t), x_{j-1}(t - \mu), \dot{x}_{j-1}(t - \mu))$$
$$x_j((j-1)\mu) = x_{j-1}((j-1)\mu)$$

for $(j-1)\mu \leq t \leq j\mu$, so that $\dot{x}(t)$ will generally be discontinuous at the values $t = k\mu$, $k \geq 0$. Note that this method is unsatisfactory when the stepsize μ is too small. Instead, asymptotic methods are then appropriate. They have been given by Vasil'eva (1962) and O'Malley (1971b). A different type of singular perturbation problem for differential–difference equations is discussed in Cooke and Meyer (1966).

A heuristic connection with familiar singular perturbation problems results if we consider the example

$$\dot{x}(t) = a\dot{x}(t - \mu), \qquad |a| < 1$$
$$x(t) = \phi(t).$$

Expanding $\dot{x}(t-\mu)$ as a power series in μ, we have

$$\dot{x}(t) = a\Big(\dot{x}(t) - \mu\ddot{x}(t) + \frac{\mu^2}{2}\dddot{x}(t) + \cdots\Big).$$

Terminating this series after two terms, note that the "truncated problem"

$$\mu a\,\ddot{x}(t) + (1-a)\,\dot{x}(t) = 0$$
$$x(0) = \phi(0), \qquad \dot{x}(0) = \dot{\phi}(0)$$

and the full problem will both have solutions converging to the constant solution $X_0(t) = \phi(0)$ of

$$\dot{X}_0(t) = a\,\dot{X}_0(t)$$
$$X_0(0) = \phi(0)$$

provided $a > 0$. Conclusions in certain applied literature involving differential–difference equations have sometimes been based on such analogies. The analogy can be misleading, however, and such conclusions should be considered dubious. For singularly perturbed ordinary differential equations, the number of initial conditions required for the full problem and the reduced problem differ by a finite number. For the differential–difference equation (4.32), however, an infinite number of initial derivatives are prescribed by (4.33) while the limiting differential equation for $X_0(t)$ requires only an initial value.

We shall seek a solution $x(t,\mu)$ of (4.32)–(4.33) of the form

$$x(t,\mu) = X(t,\mu) + \mu\, m(\theta,\mu) \tag{4.36}$$

for $t \geq 0$ where the outer solution

$$X(t,\mu) \sim \sum_{j=0}^{\infty} X_j(t)\,\mu^j$$

satisfies the difference–differential equation for $t \geq 0$ and each term m_j of the boundary layer correction

$$m(\theta,\mu) \sim \sum_{j=0}^{\infty} m_j(\theta)\,\mu^j$$

tends to zero as the boundary layer coordinate

$$\theta = \frac{t}{\mu}$$

tends to infinity. Away from $t = 0$, then, the asymptotic solution will be determined by the outer expansion. Since the initial function (4.33) is independent of μ, (4.36) implies that

$$X_j(0) = -m_{j-1}(0) \qquad \text{for each} \quad j \geq 1. \tag{4.37}$$

Thus, to calculate the outer expansion it is necessary to know the initial value $m(0, \mu)$ of the boundary layer correction.

Note that the definition of $X_0(t)$ implies that the outer expansion $X(t, \mu)$ satisfies (4.32) when $\mu = 0$. Equating coefficients of μ, X_1 must satisfy the linear differential equation

$$\begin{aligned}\dot{X}_1(t) &= f_x(t, X_0, X_0, \dot{X}_0)X_1 + f_y(t, X_0, X_0, \dot{X}_0)(X_1 - \dot{X}_0)\\ &\quad + f_u(t, X_0, X_0, \dot{X}_0)(\dot{X}_1 - \ddot{X}_0)\end{aligned}$$

or

$$\dot{X}_1(t) = A(t)X_1(t) + B_0(t),$$

where

$$\begin{aligned}A(t) &= \left(1 - f_u(t, X_0, X_0, \dot{X}_0)\right)^{-1}\\ &\quad \times \left(f_x(t, X_0, X_0, \dot{X}_0) + f_y(t, X_0, X_0, \dot{X}_0)\right)\end{aligned}$$

and $B_0(t)$ is determined from $X_0(t)$. [Note that $1 - f_u$ is invertible for $0 \leq t \leq T$ by (4.35).] In general, each coefficient $X_j(t)$ for $j \geq 1$ will be determined as a solution of a linear differential equation of the form

$$\dot{X}_j(t) = A(t)X_j(t) + B_{j-1}(t), \tag{4.38}$$

where B_{j-1} is a smooth, successively known function. Thus, the X_j's can be determined recursively on $0 \leq t \leq T$ up to specification of their initial values [by (4.37)].

The boundary layer correction terms $m_j(\theta)$ are determined successively by stepwise integration on the intervals $p \le \theta \le p+1, p \ge 0$. Since the complete expansion (4.36) and the outer expansion $X(t, \varepsilon)$ both satisfy equation (4.32), the boundary layer correction $m(\theta, \mu)$ must satisfy

$$m_\theta(\theta,\mu) = f\Big(\mu\theta, X(\mu\theta,\mu) + \mu\, m(\theta,\mu), \phi(\mu(\theta-1)), \dot{\phi}(\mu(\theta-1))\Big) - f\Big(\mu\theta, X(\mu\theta,\mu), X(\mu(\theta-1),\mu), \dot{X}(\mu(\theta-1),\mu)\Big) \quad \text{for} \quad 0 \le \theta \le 1 \tag{4.39a}$$

$$m_\theta(\theta,\mu) = f\Big(\mu\theta, X(\mu\theta,\mu) + \mu\, m(\theta,\mu), X(\mu(\theta-1),\mu) + \mu\, m(\theta-1,\mu), \dot{X}(\mu(\theta-1),\mu) + m_\theta(\theta-1,\mu)\Big) - f\Big(\mu\theta, X(\mu\theta,\mu), X(\mu(\theta-1),\mu), \dot{X}(\mu(\theta-1),\mu)\Big) \quad \text{for} \quad \theta \ge 1. \tag{4.39b}$$

For $\mu = 0$, then, $m_0(\theta)$ must be the continuous solution of

$$m_{0\theta}(\theta) = f\Big(0, X_0(0), \phi(0), \dot{\phi}(0)\Big) - f\Big(0, X_0(0), X_0(0), \dot{X}_0(0)\Big) \quad \text{for} \quad 0 \le \theta \le 1 \tag{4.40a}$$

and

$$m_{0\theta}(\theta) = f\Big(0, X_0(0), X_0(0), \dot{X}_0(0) + m_{0\theta}(\theta-1)\Big) - f\Big(0, X_0(0), X_0(0), \dot{X}_0(0)\Big) \quad \text{for} \quad \theta \ge 1. \tag{4.40b}$$

Thus, $m_{0\theta}$ is stepwise constant. Setting

$$m_{0\theta}(\theta) = G_p^0 \quad \text{for} \quad p \le \theta \le p+1, \quad p \ge 0,$$

we have $m_0(\theta) = m_0(0) + \int_0^\theta m_{0\theta}(s)\, ds$ or

$$m_0(\theta) = m_0(0) + \sum_{l=0}^{p-1} G_l^0 + (\theta - p) G_p^0 \quad \text{for} \quad p \le \theta \le p+1. \tag{4.41}$$

Since we asked that $m_0 \to 0$ as $\theta \to \infty$, we select

$$X_1(0) = -m_0(0) = \sum_{l=0}^{\infty} G_l^0, \tag{4.42}$$

assuming this limit exists. To clarify this issue, we introduce the mapping F such that

$$Fu = F^0 u = f\big(0, \phi(0), \phi(0), u\big)$$

and

$$F^j u = f\big(0, \phi(0), \phi(0), F^{j-1} u\big) \qquad \text{for each} \quad j > 1.$$

Note that (4.34) and (4.35) imply that F is a contraction mapping with $\dot{X}_0(0)$ as its (unique) fixed point [see, e.g., Hale (1969) for the definition of a fixed point and a statement of the contraction mapping principle]. Hence,

$$\dot{X}_0(0) = \lim_{p \to \infty} F^p \dot{\phi}(0)$$

and since

$$G_p^0 = F^{p+1} \dot{\phi}(0) - \dot{X}_0(0)$$

(4.42) becomes

$$X_1(0) = \lim_{p \to \infty} \left[\sum_{l=0}^{p} \left(F^{l+1} \dot{\phi}(0) \right) - (p+1) \dot{X}_0(0) \right]$$

and the limit is finite [cf. Vasil'eva (1962)]. Note that (4.40) and (4.35) imply that

$$|m_{0\theta}(\theta)| \le e^{-\kappa} |m_{0\theta}(\theta - 1)| \qquad \text{for} \quad \theta \ge 1$$

so that (estimating freely)

$$|m_{0\theta}(\theta)| \le |\dot{X}_0(0) - \dot{\phi}(0)| e^{-\kappa\theta}$$

and, by integration,

$$m_0(\theta) = O(e^{-\kappa\theta}) \qquad \text{as} \quad \theta \to \infty.$$

Equating coefficients of μ^j in (4.39) successively for each $j \geq 1$ implies the linear differential–difference equations

$$m_{j\theta}(\theta) = V_{j-1}(\theta) + W(\theta)\, m_{j\theta}(\theta - 1)$$

for $\theta \geq 0$, where V_{j-1} is a linear combination of $m_l(\theta)$, $m_l(\theta - 1)$, and $m_{l\theta}(\theta - 1)$ for $l < j$ and

$$W(\theta) = \begin{cases} 0 & \text{for} \quad 0 \leq \theta \leq 1 \\ f_u\big(0, \phi(0), \phi(0), \dot{X}_0(0) + m_{0\theta}(\theta - 1)\big) & \text{for} \quad \theta \geq 1. \end{cases}$$

Thus, we have

$$m_{j\theta}(\theta) = G_p^j(\theta) \qquad \text{for} \quad p \leq \theta \leq p + 1$$

and the G_p^j's are determined, in turn, for $p = 0, 1, 2, \ldots$. Integrating stepwise, we have

$$m_j(\theta) = m_j(0) + \sum_{l=0}^{p-1} \left(\int_l^{l+1} G_l^j(s)\, ds \right) + \int_p^\theta G_p^j(s)\, ds$$

$$\text{for} \quad p \leq \theta \leq p + 1. \tag{4.43}$$

Note that the derivatives of the m_j's will generally be discontinuous at integer values of θ.

Since $m_j \to 0$ as $\theta \to \infty$, we must select

$$X_{j+1}(0) = -m_j(0) = \sum_{l=0}^{\infty} \left(\int_l^{l+1} G_l^j(s)\, ds \right). \tag{4.44}$$

Further, since (by induction)

$$V_{j-1}(\theta) = O(e^{-\kappa(1-\delta)\theta}) \qquad \text{as} \quad \theta \to \infty,$$

integration implies that $m_j(\theta)$ is also exponentially decaying as $\theta \to \infty$ and that the sum in (4.44) is finite.

Summarizing, we have

THEOREM 4: *Under the assumptions* (Hi)–(Hiii), *the initial value problem* (4.32)–(4.33) *has a unique solution* $x(t, \mu)$ *for* μ *sufficiently small which is such that, for each integer* $N \geq 0$,

$$x(t, \mu) = X_0(t) + \sum_{j=1}^{N} \left(X_j(t) + m_{j-1}(t/\mu)\right)\mu^j + \mu^{N+1} R(t, \mu),$$

where the m_j*'s* $\to 0$ *as* $t/\mu \to \infty$ *and* $R(t, \mu)$ *is uniformly bounded throughout* $0 \leq t \leq T$.

The theorem can be proved as has been done by O'Malley (1971b). No further complications arise if the functions f and ϕ have asymptotic expansions as $\mu \to 0$. We note that the steplike construction of the boundary layer correction (cf. Fig. 14) is quite novel compared to the preceding, but that the result is still exponentially decaying as $\theta \to \infty$.

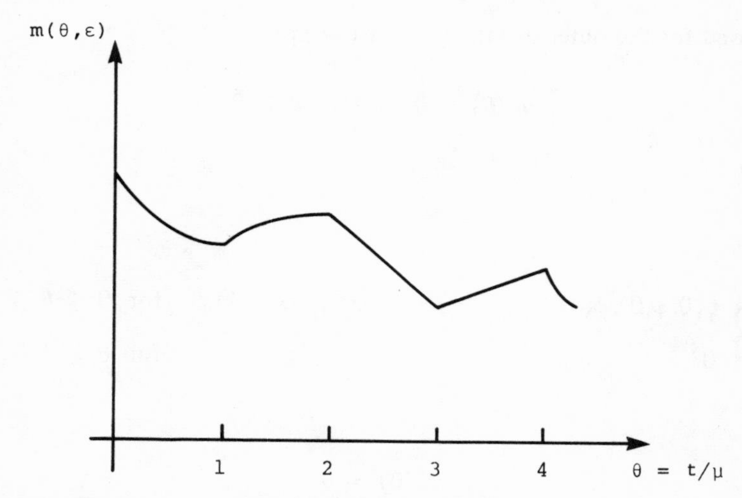

FIGURE 14 The steplike nature of the boundary layer correction for the differential–difference problem (4.32)–(4.33).

When $f_u \equiv 0$, Eq. (4.32) is called an equation with retarded argument. For such problems, calculation of the boundary layer correction terms becomes considerably simplified. In particular, one has

$$m_j(\theta) \equiv 0 \qquad \text{for} \quad \theta \geq j;$$

i.e., these terms ultimately become zero. Here $m_{j\theta} = V_{j-1}$ is known successively, and we have

$$m_j(\theta) = \begin{cases} -\int_\theta^j V_{j-1}(s)\,ds & \text{for} \quad \theta \leq j \\ 0 & \text{for} \quad \theta \geq j. \end{cases}$$

This implies the initial value

$$X_j(0) = \int_0^{j-1} V_{j-2}(s)\,ds$$

needed for the outer expansion. For example,

$$m_0(\theta) \equiv 0 \qquad \text{for} \quad \theta \geq 0$$

and

$$m_1(\theta)$$

$$= \begin{cases} f_y\big(0, \phi(0), \phi(0), 0\big)\big(\dot{X}_0(0) - \dot{\phi}(0)\big)\int_\theta^1 (s-1)\,ds & \text{for} \ \ 0 \leq \theta \leq 1 \\ 0 & \text{for} \ \ \theta \geq 1 \end{cases}$$

so

$$X_1(0) = 0$$

and

$$X_2(0) = \tfrac{1}{2} f_y\big(0, \phi(0), \phi(0), 0\big)\big(\dot{X}_0(0) - \dot{\phi}(0)\big).$$

Since $m_j(\theta) = 0$ for $\theta \geq j$, Theorem 4 implies

COROLLARY: *Consider the initial value problem* (4.32)–(4.33) *when f is independent of u and assumptions* (Hi) *and* (Hii) *hold. Then, for* μ *sufficiently small, there is a unique solution* $x(t, \mu)$ *for* $0 \leq t \leq T$ *which is such that, for each integer* $N \geq 1$,

$$x(t, \mu) = \sum_{j=0}^{N} X_j(t)\mu^j + O(\mu^{N+1}) \qquad \text{for} \quad \mu(N-1) \leq t \leq T.$$

We note that equations with small retarded arguments occur in many applications. For example, they have often been used in population models to improve on the classical Volterra–Lotka models which involve no delay. There, the biologist Hutchinson (1948) states "there is a tendency for the time lag to be reduced as much as possible by natural selection." Thus arguments for small delay problems are found throughout the literature on epidemics and population. An interesting application of these methods to an optimal control problem is given by Sannuti and Reddy (1973).

EXAMPLE: Let us examine in detail a special problem where we have a linear difference equation in $\dot{x}(t)$, i.e.,

$$\begin{aligned} \dot{x}(t) &= a\dot{x}(t-\mu), & t &\geq 0 \\ x(t) &= \phi(t), & -\mu &\leq t \leq 0 \end{aligned}$$

for $|a| < 1$. Here the reduced problem

$$\begin{aligned} \dot{x}(t) &= a\dot{x}(t) \\ x(0) &= \phi(0) \end{aligned}$$

has the unique solution $X_0(t) = \phi(0)$. Further, if we seek an outer expansion

$$X(t, \mu) \sim \phi(0) + \mu X_1(t) + \mu^2 X_2(t) + \mu^3(\cdots)$$

for $t > 0$, we formally have

$$\dot{X}(t, \mu) \sim \mu \dot{X}_1(t) + \mu^2 \dot{X}_2(t) + \mu^3(\cdots)$$

and

$$\dot{X}(t - \mu, \mu) \sim \mu \dot{X}_1(t) + \mu^2 \big(\dot{X}_2(t) - \ddot{X}_1(t)\big) + \mu^3(\cdots).$$

Equating coefficients, then, in the difference equation

$$\dot{X}(t, \mu) = a\dot{X}(t - \mu, \mu)$$

implies

$$\dot{X}_1(t) = a\dot{X}_1(t)$$
$$\dot{X}_2(t) = a\big(\dot{X}_2(t) - \ddot{X}_1(t)\big),$$

etc. Thus, each $X_j(t)$ is constant and the values $X_j(0)$ will be determined through the boundary layer correction terms. Moreover, since the outer solution is a constant function of t, a boundary layer correction at $t = 0$ will be clearly necessary unless $\phi(t) = \phi(0)$.

Representing the solution in the form

$$x(t, \mu) = X(0, \mu) + \mu m(\theta, \mu)$$

for $t \geq 0$ and $\theta = t/\mu$ [since $X(t, \mu) = X(0, \mu)$], the boundary layer correction $m(\theta, \mu)$ must satisfy the difference equation

$$m_\theta(\theta, \mu) = \begin{cases} a\dot{\phi}\big(\mu(\theta - 1)\big), & 0 \leq \theta \leq 1 \\ am_\theta(\theta - 1, \mu), & \theta \geq 1. \end{cases}$$

Thus,

$$m_\theta(\theta, \mu) = a^{p+1}\dot{\phi}\big(\mu(\theta - p - 1)\big), \qquad p \leq \theta \leq p + 1, \quad p \geq 0.$$

Integrating, then,

$$m(\theta, \mu) = m(0, \mu) + \sum_{k=0}^{p} \left(\int_{k}^{k+1} m_\theta(s, \mu)\, ds \right) + \int_{p}^{\theta} m_\theta(s, \mu)\, ds$$

$$\text{for} \quad p \leq \theta \leq p + 1$$

and since m must tend to zero as $\theta \to \infty$, we select

$$m(0, \mu) = -\lim_{p \to \infty} \left(\sum_{k=0}^{p} a^{k+1} \right) \int_{-1}^{0} \dot{\phi}(\mu s)\, ds$$

$$= \frac{1}{\mu}\left(\frac{a}{1-a} \right)\big(\phi(-\mu) - \phi(0)\big).$$

Thus,

$$m(\theta, \mu) = \frac{a^{p+1}}{\mu}\left[\frac{\phi(-\mu) - \phi(0)}{1-a} + \phi\big(\mu(\theta - p - 1)\big) - \phi(-\mu) \right]$$

$$= \frac{a^{p+1}}{\mu(1-a)}[a\,\phi(-\mu) + (1-a)\,\phi(\mu(\theta - p - 1)) - \phi(0)]$$

$$\text{for} \quad p \leq \theta \leq p + 1, \quad p \geq 0.$$

In particular, it is important to note that m decays like $e^{-\kappa\theta}$ as $\theta \to \infty$ for $\kappa = -\ln|a|$.

Finally, the initial condition implies $\phi(0) = X(0, \mu) + \mu m(0, \mu)$, so we have obtained the outer expansion as

$$X(t, \mu) = X(0, \mu) = \frac{1}{1-a}[a\,\phi(-\mu) + (1-2a)\,\phi(0)].$$

Thus, the asymptotic solution for $t > 0$ is given by the outer expansion with

$$X_j(t) = \left(\frac{a}{1-a} \right)(-1)^j \frac{\phi^{(j)}(0)}{j!}, \qquad j > 0.$$

Near $t = 0$, however, the boundary layer correction is important. It is given by $\mu m(t/\mu, \mu) \sim \mu \sum_{j=0}^{\infty} m_j(t/\mu)\mu^j$, where

$$m_j(\theta) = \frac{a^{p+1}}{1-a}\frac{\phi^{(j+1)}(0)}{(j+1)!}\left(a(-1)^j + (1-a)(\theta - p - 1)^{j+1}\right),$$

$$p \leq \theta \leq p+1, \quad p \geq 0.$$

Alternatively, using the greatest integer function $[\cdot]$, we have a uniformly valid expansion for $t \geq 0$ given by

$$x(t, \mu) \sim \phi(0) + \left(\frac{a}{1-a}\right)\sum_{j=1}^{\infty}\frac{\phi^{(j)}(0)(-\mu)^j}{j!}$$

$$\times\left(1 + a^{[t/\mu]}\left\{-a + (1-a)\left(1 - \frac{t}{\mu} + \left[\frac{t}{\mu}\right]\right)^{j+1}\right\}\right).$$

CHAPTER 5

NONLINEAR BOUNDARY VALUE PROBLEMS

1. SOME SECOND-ORDER SCALAR PROBLEMS

We would now like to consider three examples of boundary value problems of the form

$$\begin{gathered}\varepsilon\frac{d^2x}{dt^2} = F\left(x, \frac{dx}{dt}, t, \varepsilon\right), \qquad 0 \le t \le 1 \\ x(0,\varepsilon) = \alpha(\varepsilon), \qquad x(1,\varepsilon) = \beta(\varepsilon)\end{gathered} \tag{5.1}$$

which illustrate the varied phenomena possible in the limit as $\varepsilon \to 0$. In Section 2, we will obtain asymptotic solutions for a restricted class of problems.

EXAMPLE 1: Coddington and Levinson (1952) introduced the example

$$\varepsilon\frac{d^2x}{dt^2} + \frac{dx}{dt} + \left(\frac{dx}{dt}\right)^3 = 0$$
$$x(0) = \alpha, \qquad x(1) = \beta. \tag{5.2}$$

Here the differential equation can be explicitly integrated (since it is a Bernoulli equation in dx/dt). Applying the boundary conditions, however, one finds that the problem has no solution for ε sufficiently small when the constants α and β are unequal.

As this example would suggest, progress toward obtaining asymptotic solutions to problems like (5.1), where $F(x, dx/dt, t, 0)$ is nonlinear in dx/dt, has been limited. Vishik and Lyusternik (1958, 1960) have discussed the general case where $F(x, dx/dt, t, 0) = O((dx/dt)^l)$, $l > 0$, as $|dx/dt| \to \infty$. They conclude, under appropriate assumptions, that $x'(0)$ will remain bounded as $\varepsilon \to 0$ and nonuniform convergence as $\varepsilon \to 0$ cannot occur at $t = 0$ if $l > 2$. (Note that $l = 3$ in Coddington and Levinson's example.) However, it is possible to have

$$x'(0) = O(\varepsilon^{-1/(2-l)}) \qquad \text{if} \quad 0 < l < 2$$

and

$$x'(0) = O(e^{C/\varepsilon}) \qquad \text{for some} \quad C > 0 \quad \text{if} \quad l = 2.$$

The same conclusions hold at the endpoint $t = 1$. We note that these estimates on endpoint behavior could be helpful in selecting appropriate stretched variables (or boundary layer coordinates). Vishik and Lyusternik also give complicated expansion techniques for such problems which have been generalized to systems by Kasymov (1968). Similar problems are also discussed by Yarmish (1972). Cohen (1973b) obtains the limiting behavior for a restricted class of nonlinear problems.

EXAMPLE 2: The nonlinear problem

$$\varepsilon\frac{d^2x}{dt^2} + \frac{dx}{dt} - \left(\frac{dx}{dt}\right)^2 = 0$$
$$x(0) = 1, \qquad x(1) = 0 \tag{5.3}$$

has the solution

$$x(t,\varepsilon) = -\varepsilon\ln[1 + e^{-1/\varepsilon} - e^{-t/\varepsilon}]$$

so $x \to 0$ as $\varepsilon \to 0$ away from $t = 0$. Further

$$\frac{dx}{dt} = -e^{-t/\varepsilon}[1 + e^{-1/\varepsilon} - e^{-t/\varepsilon}]^{-1}$$

so $x'(0) = -e^{1/\varepsilon}$, as Vishik and Lyusternik's results anticipate.

Other problems produce nonuniformities in the interior of the interval [0, 1]. Thus, consider

EXAMPLE 3:

$$\varepsilon\frac{d^2x}{dt^2} - \frac{dx}{dt} + \left(\frac{dx}{dt}\right)^3 = 0$$
$$x(0) = 0, \qquad x(1) = \tfrac{1}{2}. \tag{5.4}$$

We note that this equation differs from that of Coddington and Levinson only in the sign of the "damping" term. The solution here, however, is defined as $\varepsilon \to 0$ and satisfies

$$x(t,\varepsilon) = \varepsilon\log\Big[\tfrac{1}{2}e^{(t-1/2)/\varepsilon} + [1 + \tfrac{1}{4}e^{2(t-1/2)/\varepsilon} + O(e^{-1/2\varepsilon})]^{1/2}\Big] + O(\varepsilon e^{-1/2\varepsilon}).$$

Thus, the limiting solution as $\varepsilon \to 0$ (cf. Fig. 15) is

$$X_0(t) = \begin{cases} 0 & \text{for } 0 \le t \le 1/2 \\ t - 1/2 & \text{for } 1/2 \le t \le 1. \end{cases}$$

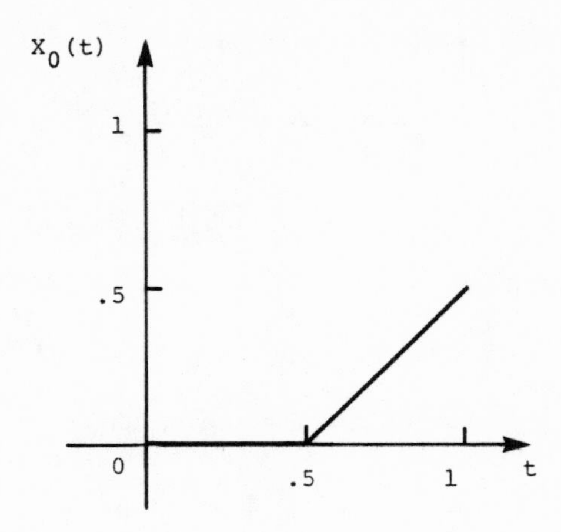

FIGURE 15 The limiting solution for $\varepsilon\ddot{x} - \dot{x} + (\dot{x})^3 = 0$, $x(0) = 0$, $x(1) = \frac{1}{2}$, $\varepsilon \to 0$.

Note that this limiting solution satisfies the reduced equation throughout $[0, 1]$ but has a discontinuous derivative at $t = \frac{1}{2}$. We note that other such "solutions" of the reduced equation satisfying the boundary conditions are given by

$$\tilde{X}(t) = \begin{cases} t, & 0 \leq t \leq \frac{1}{2} \\ \frac{1}{2}, & \frac{1}{2} \leq t \leq 1, \end{cases}$$

$$\tilde{\tilde{X}}(t) = \begin{cases} -t, & 0 \leq t \leq \frac{1}{4} \\ \frac{1}{4} + t, & \frac{1}{4} \leq t \leq 1, \end{cases}$$

and

$$\tilde{\tilde{\tilde{X}}}(t) = \begin{cases} t, & 0 \leq t \leq \frac{3}{4} \\ \frac{3}{4} - t, & \frac{3}{4} \leq t \leq 1. \end{cases}$$

These, however, do not provide limiting solutions for the original problem as $\varepsilon \to 0$.

Such "angular" limiting solutions result for a class of nonlinear equations [cf. Haber and Levinson (1955)]. Further, asymptotic expansions of such solutions can be obtained by the same techniques used in Chapter 4 for initial value problems [cf. O'Malley (1970b)]. Another type of nonlinear problem with angular solutions is typified by Example 4 of Section 1.1, while the "transition layer" problems of Fife (1974) represent another type of interior nonuniformity.

2. SECOND-ORDER QUASI-LINEAR EQUATIONS

In this section, we shall consider the quasi-linear problem

$$\begin{aligned} &\varepsilon x'' + f(t,x)x' + g(t,x) = 0 \\ &x(0,\varepsilon) = \alpha, \qquad x(1,\varepsilon) = \beta \end{aligned} \tag{5.5}$$

for prescribed constants α and β under the hypotheses that

(Hi) *the terminal value problem*

$$\begin{aligned} &f(t,x)x' + g(t,x) = 0 \\ &x(1) = \beta \end{aligned} \tag{5.6}$$

has a solution $X_0(t)$ *for* $0 \leq t \leq 1$ *such that, for some* $\kappa > 0$,

$$f\big(t, X_0(t)\big) \geq \kappa \tag{5.7}$$

there, and that

(Hii)

$$f(0,\chi) \geq \kappa \tag{5.8}$$

for all χ *between* α *and* $X_0(0)$.

We will also suppose that $f(t, x)$ and $g(t, x)$ are infinitely differentiable in the indicated domains. We note that this problem has been discussed by several authors, including Coddington and Levinson (1952), Wasow (1956), Willett (1966), Erdélyi (1968a), and O'Malley (1968a). Recall that the linear problem where $f(t, x) = a(t)$ and $g(t, x) = b(t)x$ was solved (by methods quite different than below) in Section 3.1.

Under the hypotheses (Hi) and (Hii), we shall obtain an asymptotic solution to (5.5) of the form

$$x(t, \varepsilon) = X(t, \varepsilon) + \xi(\tau, \varepsilon), \tag{5.9}$$

where

$$X(t, \varepsilon) \sim \sum_{j=0}^{\infty} X_j(t)\varepsilon^j$$

$$\xi(\tau, \varepsilon) \sim \sum_{j=0}^{\infty} \xi_j(\tau)\varepsilon^j$$

as $\varepsilon \to 0$ and τ is the stretched variable

$$\tau = t/\varepsilon.$$

Here the terms ξ_j and their derivatives $d\xi_j/d\tau$ should tend to zero as $\tau \to \infty$. Away from $t = 0$, then, we will have $x(t, \varepsilon) \to X(t, \varepsilon)$ as $\varepsilon \to 0$.

Proceeding formally, we must ask that the asymptotic expansion of the outer solution $X(t, \varepsilon)$ satisfy the differential equation of (5.5) for $t > 0$. Since the first term $X_0(t)$ satisfies the nonlinear reduced problem (5.6), the second term X_1 must satisfy the linear equation

$$f(t, X_0)X_1' + \big(f_x(t, X_0)X_0' + g_x(t, X_0)\big)X_1 = -X_0'',$$

and higher terms X_j will satisfy analogous differential equations with only a different, but successively known right-hand side. Since the boundary layer correction $\xi(\tau, \varepsilon)$ is asymptotically negligible at $t = 1$, we must have $X(1, \varepsilon) = \beta$. Hence $X_j(1) = 0$ for each $j \geq 1$ and further terms in the outer expansion can be uniquely determined successively from their linear differential equations. Thus, the asymptotic solution away from $t = 0$ can be formally determined without use of a boundary layer correction. Note that this contrasts with the situation for initial value problems where the outer expansion cannot be obtained without determining the initial values of the boundary layer correction.

Since the outer solution satisfies (5.5), the boundary layer correction $\xi(\tau, \varepsilon)$ must satisfy the nonlinear equation

$$\begin{aligned} &\frac{d^2\xi}{d\tau^2} + f\Big(\varepsilon\tau, X(\varepsilon\tau, \varepsilon) + \xi(\tau, \varepsilon)\Big)\frac{d\xi}{d\tau} \\ &\quad + \varepsilon\Big[\Big(f(\varepsilon\tau, X(\varepsilon\tau, \varepsilon) + \xi(\tau, \varepsilon)) - f(\varepsilon\tau, X(\varepsilon\tau, \varepsilon))\Big)\frac{dX}{dt} \\ &\quad + g\Big(\varepsilon\tau, X(\varepsilon\tau, \varepsilon) + \xi(\tau, \varepsilon)\Big) - g\Big(\varepsilon\tau, X(\varepsilon\tau, \varepsilon)\Big)\Big] = 0 \qquad (5.10) \end{aligned}$$

and the initial condition

$$\xi(0, \varepsilon) = \alpha - X(0, \varepsilon).$$

When $\varepsilon = 0$, then, we have

$$\frac{d^2\xi_0}{d\tau^2} + f\Big(0, X_0(0) + \xi_0(\tau)\Big)\frac{d\xi_0}{d\tau} = 0.$$

Since both ξ_0 and $d\xi_0/d\tau \to 0$ as $\tau \to \infty$, ξ_0 must satisfy the nonlinear initial value problem

$$\begin{aligned} &\frac{d\xi_0}{d\tau} = -\int_0^{\xi_0} f\Big(0, X_0(0) + r\Big)\, dr \\ &\xi_0(0) = \alpha - X_0(0) \end{aligned} \qquad (5.11)$$

on the interval $\tau \geq 0$. Because f remains positive, $|\xi_0(\tau)|$ decreases monotonically as τ increases. Thus, ξ_0 exists for all $\tau \geq 0$ and, by (5.8), satisfies

$$|\xi_0(\tau)| \leq |\xi_0(0)|e^{-\kappa\tau} \qquad \text{for} \quad \tau \geq 0.$$

In general, of course, we cannot obtain an explicit solution of (5.11), but we can obtain the unique solution through successive approximations.

Further terms of the boundary layer correction, however, satisfy linear equations

$$\frac{d^2\xi_j}{d\tau^2} + f\Big(0, X_0(0) + \xi_0(\tau)\Big)\frac{d\xi_j}{d\tau} + f_x\Big(0, X_0(0) + \xi_0(\tau)\Big)\frac{d\xi_0}{d\tau}\xi_j = C_{j-1}(\tau),$$

where the C_{j-1}'s are known successively and are $O(e^{-\kappa(1-\delta)\tau})$ as $\tau \to \infty$ for any $\delta > 0$. Direct integration then implies that

$$\begin{aligned}\xi_j(\tau) = &-X_j(0)\exp\left[-\int_0^\tau f\Big(0, X_0(0) + \xi_0(s)\Big)\, ds\right] \\ &- \int_0^\tau \exp\left[-\int_p^\tau f\Big(0, X_0(0) + \xi_0(s)\Big)\, ds\right] \\ &\times \int_p^\infty C_{j-1}(r)\, dr\, dp \end{aligned} \tag{5.12}$$

and, since $f\Big(0, X_0(0) + \xi_0(\tau)\Big) \geq \kappa$ for all $\tau \geq 0$,

$$\xi_j(\tau) = O(e^{-\kappa(1-\delta)\tau}) \qquad \text{as} \quad \tau \to \infty.$$

Formally, then, we have uniquely obtained an asymptotic solution $x(t, \varepsilon)$ of the boundary value problem (5.5). For any integer $N \geq 0$, let

$$x(t, \varepsilon) = \sum_{j=0}^{N} \Big(X_j(t) + \xi_j(t/\varepsilon)\Big)\varepsilon^j + \varepsilon^{N+1}R(t, \varepsilon). \tag{5.13}$$

Then it can be shown [cf. O'Malley (1969a)] that, for ε sufficiently small, the boundary value problem has a unique solution which is of the form (5.13) with $R(t, \varepsilon)$ uniformly bounded throughout $0 \leq t \leq 1$. If $f(t, x)$ were negative, we would instead have nonuniform convergence at $t = 1$ with the appropriately altered hypotheses. Note that hypothesis (Hii) might be weakened to the condition that

$$\frac{1}{\xi} \int_0^\xi f\big(0, X_0(0) + r\big)\, dr \geq \kappa$$

for all ξ values between $\alpha - X_0(0)$ and 0 [cf. Fife (1973)].

As an example, consider the highly nonlinear problem

$$\varepsilon x'' + e^x x' - \left(\frac{\pi}{2} \sin \frac{\pi t}{2}\right) e^{2x} = 0$$

$$x(0) = \alpha, \qquad x(1) = 0.$$

The reduced problem

$$e^x x' - \left(\frac{\pi}{2} \sin \frac{\pi t}{2}\right) e^{2x} = 0, \qquad x(1) = 0$$

has the solution

$$X_0(t) = -\ln\left(1 + \cos \frac{\pi t}{2}\right), \qquad 0 \leq t \leq 1.$$

The first term ξ_0 of the boundary layer correction must satisfy

$$\frac{d\xi_0}{d\tau} = -\int_0^{\xi_0} e^{X_0(0)} e^r\, dr, \qquad \xi_0(0) = \alpha + \ln 2.$$

Thus,

$$\xi_0(\tau) = -\ln[(1 - e^{-\tau/2}) + \tfrac{1}{2} e^{-\alpha} e^{-\tau/2}] = O(e^{-\tau/2}),$$

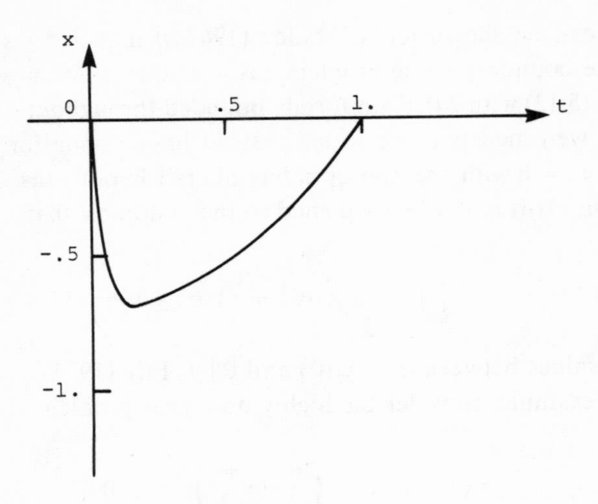

FIGURE 16 The limiting solution of $\varepsilon x'' + e^x x' - (\pi/2)\sin \pi t/2 = 0$, $x(0) = x(1) = 0$.

so

$$x(t, \varepsilon) = -\ln\left[\left(1 + \cos\frac{\pi t}{2}\right)\left(1 - e^{-t/2\varepsilon} + \frac{e^{-\alpha}}{2}e^{-t/2\varepsilon}\right)\right] + O(\varepsilon) \qquad \text{for} \quad 0 \leq t \leq 1$$

(cf. Fig. 16).

Note that this method can also be used to obtain the asymptotic solution of Example 4, Section 1.1, in Cases 1 and 2. It does not apply, however, in Cases 3–6.

Before proceeding, we wish to observe that the expansion procedure developed here for $\varepsilon x'' + f(t,x)x' + g(t,x) = 0$ will not work when $f(t,x)$ has a zero. In the special case, however, when $f(t,x) \equiv 0$, i.e.,

$$\varepsilon x'' + g(t,x) = 0$$
$$x(0) = \alpha, \qquad x(1) = \beta,$$

a limiting solution within (0, 1) will be given by any $X_0(t)$ such that

$$g\big(t, X_0(t)\big) = 0$$

provided $g_x(t, X_0(t)) < 0$ and $g_x(0,x)$ and $g_x(1,x)$ are appropriately restricted [cf. Vasil'eva and Tupciev (1960), Yarmish (1972), and Fife (1973)]. At $t = 0$, nonuniform convergence will generally require a boundary layer correction in the stretched variable $\tau_0 = t/\sqrt{\varepsilon}$, while at $t = 1$ there will be nonuniform convergence involving the variable $\tau_1 = (1 - t)/\sqrt{\varepsilon}$.

3. QUASI-LINEAR SYSTEMS

Generalizing the boundary value problem (5.5), let us consider the quasi-linear system

$$\begin{aligned} \frac{dx}{dt} &= f(x,y,t,\varepsilon) \equiv f_1(x,t,\varepsilon) + f_2(x,t,\varepsilon)y \\ \varepsilon\frac{dy}{dt} &= g(x,y,t,\varepsilon) \equiv g_1(x,t,\varepsilon) + g_2(x,t,\varepsilon)y \end{aligned} \tag{5.14}$$

of two scalar equations on the interval $0 \le t \le 1$ with the boundary conditions

$$\begin{aligned} a_1(\varepsilon)x(0,\varepsilon) + \varepsilon a_2(\varepsilon)y(0,\varepsilon) &= \alpha(\varepsilon) \\ b_1(\varepsilon)x(1,\varepsilon) + b_2(\varepsilon)y(1,\varepsilon) &= \beta(\varepsilon), \end{aligned} \tag{5.15}$$

where the f_i, g_i, a_i, b_i, α, and β all have asymptotic series expansions as $\varepsilon \to 0$ such that the coefficients in the expansions for f_i and g_i are infinitely differentiable functions of x and t. (A reason for introducing the ε multiple of a_2 will be given below.)

We note that the scalar problem (5.5) can be put in the form (5.14)–(5.15) as follows:

$$\begin{aligned} &\frac{dx}{dt} = y \\ &\varepsilon\frac{dy}{dt} = -g(t,x) - f(t,x)y \\ &x(0,\varepsilon) = \alpha, \qquad x(1,\varepsilon) = \beta. \end{aligned} \tag{5.16}$$

For f positive, recall that the limiting solution as $\varepsilon \to 0$ features nonuniform convergence at $t = 0$ such that x is bounded and y is unbounded there.

By analogy, we shall seek conditions which imply convergence of the solution of the problem (5.14)–(5.15) away from $t = 0$ to a solution of the reduced problem

$$\begin{aligned} \frac{dx}{dt} &= f_1(x,t,0) + f_2(x,t,0)y \\ 0 &= g_1(x,t,0) + g_2(x,t,0)y \\ b_1(0)x(1) + b_2(0)y(1) &= \beta(0) \end{aligned}$$

as $\varepsilon \to 0$. To satisfy this problem, we would need

$$y = -\frac{g_1(x,t,0)}{g_2(x,t,0)}$$

and

$$\frac{dx}{dt} = f_1(x,t,0) - f_2(x,t,0)\,\frac{g_1(x,t,0)}{g_2(x,t,0)},$$

where the terminal value $x(1)$ is a root γ of the nonlinear equation

$$P(\gamma) = b_1(0)\gamma - b_2(0)\,\frac{g_1(\gamma,1,0)}{g_2(\gamma,1,0)} - \beta(0) = 0.$$

Selecting some root γ, define $X_0(t)$ and $Y_0(t)$ by asking that X_0 satisfy the terminal value problem

$$\begin{aligned} \frac{dX_0}{dt} &= f_1(X_0,t,0) - f_2(X_0,t,0)\,\frac{g_1(X_0,t,0)}{g_2(X_0,t,0)} \\ X_0(1) &= \gamma \end{aligned} \tag{5.17}$$

with

$$Y_0(t) = -\frac{g_1(X_0,t,0)}{g_2(X_0,t,0)}.$$

Then (X_0, Y_0) will satisfy the reduced problem.

Let us assume that

(Hi) *a solution* $(X_0(t), Y_0(t))$ *of* (5.17) *is defined on* $0 \leq t \leq 1$ *and is such that*

$$g_2\big(X_0(t), t, 0\big) \leq -\kappa \tag{5.18}$$

there (*for some* $\kappa > 0$) *and that*

$$b_1(0) - b_2(0)\,\frac{g_x\big(X_0(1), Y_0(1), 1, 0\big)}{g_2\big(X_0(1), 1, 0\big)} \neq 0. \tag{5.19}$$

Note that such a solution of the reduced problem will be uniquely determined once $X_0(1)$ is selected and that (5.19) implies that $X_0(1)$ is a simple root of $P(\gamma) = 0$, i.e., $P'(\gamma) \neq 0$. [Recall that for the problem (5.16), (5.18) requires the reduced problem to have a solution on $0 \leq t \leq 1$ such that the positivity condition (5.7) holds. For (5.16), (5.19) is always satisfied.]

We shall now seek an asymptotic solution of (5.14)–(5.15) of the form

$$\begin{aligned} x(t, \varepsilon) &= X(t, \varepsilon) + \xi(\tau, \varepsilon) \\ y(t, \varepsilon) &= Y(t, \varepsilon) + \frac{1}{\varepsilon}\eta(\tau, \varepsilon), \end{aligned} \tag{5.20}$$

where

$$X(t, \varepsilon) \sim \sum_{j=0}^{\infty} X_j(t)\varepsilon^j, \qquad Y(t, \varepsilon) \sim \sum_{j=0}^{\infty} Y_j(t)\varepsilon^j$$

$$\xi(\tau, \varepsilon) \sim \sum_{j=0}^{\infty} \xi_j(\tau)\varepsilon^j, \qquad \eta(\tau, \varepsilon) \sim \sum_{j=0}^{\infty} \eta_j(\tau)\varepsilon^j$$

as $\varepsilon \to 0$. Here (X_0, Y_0) is the solution of the reduced problem under consideration and the terms ξ_j and η_j all tend to zero as the stretched variable $\tau = t/\varepsilon$ tends to infinity. Away from $t = 0$, then, (x, y) will be asymptotically represented by the outer solution $(X(t,\varepsilon), Y(t,\varepsilon))$. The boundary layer correction $(\xi(\tau,\varepsilon), \eta(\tau,\varepsilon)/\varepsilon)$ is needed to obtain the nonuniform convergence at $t = 0$. Note, in particular, that we anticipate that y will generally be unbounded at $t = 0$. We shall proceed to formally generate the expansions, giving the additional hypothesis (Hii) when needed to construct the boundary layer correction.

The outer solution $(X(t,\varepsilon), Y(t,\varepsilon))$ must satisfy the system (5.14) and the terminal boundary condition. When $\varepsilon = 0$, these equations are satisfied by the solution $(X_0(t), Y_0(t))$ of the reduced problem. Higher-order coefficients in the expansion must therefore satisfy linear systems of the form

$$\frac{dX_j}{dt} = f_x\big(X_0(t), Y_0(t), t, 0\big)X_j + f_2\big(X_0(t), t, 0\big)Y_j + P_{j-1}(t)$$
$$0 = g_x\big(X_0(t), Y_0(t), t, 0\big)X_j + g_2\big(X_0(t), t, 0\big)Y_j + Q_{j-1}(t)$$

on $0 \leq t \leq 1$ plus the boundary condition

$$b_1(0)\,X_j(1) + b_2(0)\,Y_j(1) = \tilde{\beta}_{j-1},$$

where P_{j-1}, Q_{j-1}, and $\tilde{\beta}_{j-1}$ are known successively in terms of preceding coefficients. By assumption (5.18), $Y_j(t)$ is determined as a linear function of $X_j(t)$; i.e.,

$$Y_j(t) = \frac{-[g_x(X_0, Y_0, t)\,X_j(t) + Q_{j-1}(t)]}{g_2(X_0, t, 0)}.$$

Thus

$$\left[b_1(0) - b_2(0)\frac{g_x\big(X_0(1), Y_0(1), 1, 0\big)}{g_2\big(X_0(1), 1, 0\big)}\right]X_j(1)$$

is known and assumption (5.19) implies that the terminal value $X_j(1)$ is uniquely determined. Finally, integrating the resulting linear terminal value problem for $X_j(t)$ determines both X_j and Y_j uniquely throughout $0 \leq t \leq 1$. Assumption (Hi), then, suffices to formally obtain the outer expansion (X, Y) uniquely. Note that no knowledge of the boundary layer correction terms is required to determine this outer expansion, which hopefully provides an asymptotic solution away from $t = 0$.

Since the outer solution satisfies the system (5.14), (5.20) implies that the boundary layer correction must satisfy the nonlinear system

$$\begin{aligned}\frac{d\xi}{d\tau} = {} & \eta(\tau,\varepsilon) f_2\big(X(\varepsilon\tau,\varepsilon) + \xi(\tau,\varepsilon), \varepsilon\tau, \varepsilon\big) \\ & + \varepsilon\Big[f_1\big(X(\varepsilon\tau,\varepsilon) + \xi(\tau,\varepsilon), \varepsilon\tau, \varepsilon\big) - f_1\big(X(\varepsilon\tau,\varepsilon), \varepsilon\tau, \varepsilon\big) \\ & + Y(\varepsilon\tau,\varepsilon)\Big(f_2(X(\varepsilon\tau,\varepsilon) + \xi(\tau,\varepsilon), \varepsilon\tau, \varepsilon) \\ & - f_2(X(\varepsilon\tau,\varepsilon), \varepsilon\tau, \varepsilon)\Big)\Big] \end{aligned} \tag{5.21a}$$

$$\begin{aligned}\frac{d\eta}{d\tau} = {} & \eta(\tau,\varepsilon) g_2\big(X(\varepsilon\tau,\varepsilon) + \xi(\tau,\varepsilon), \varepsilon\tau, \varepsilon\big) \\ & + \varepsilon\Big[g_1\big(X(\varepsilon\tau,\varepsilon) + \xi(\tau,\varepsilon), \varepsilon\tau, \varepsilon\big) - g_1\big(X(\varepsilon\tau,\varepsilon), \varepsilon\tau, \varepsilon\big) \\ & + Y(\varepsilon\tau,\varepsilon)\Big(g_2(X(\varepsilon\tau,\varepsilon) + \xi(\tau,\varepsilon), \varepsilon\tau, \varepsilon) - g_2(X(\varepsilon\tau,\varepsilon), \varepsilon\tau, \varepsilon)\Big)\Big] \end{aligned} \tag{5.21b}$$

as well as the initial condition

$$a_1(\varepsilon)\xi(0,\varepsilon) + a_2(\varepsilon)\eta(0,\varepsilon) = \alpha(\varepsilon) - a_1(\varepsilon)X(0,\varepsilon) - \varepsilon a_2(\varepsilon)Y(0,\varepsilon). \tag{5.22}$$

At $\varepsilon = 0$, then, we ask that

$$\begin{aligned}\frac{d\xi_0}{d\tau} &= \eta_0(\tau) f_2\big(X_0(0) + \xi_0(\tau), 0, 0\big) \\ \frac{d\eta_0}{d\tau} &= \eta_0(\tau) g_2\big(X_0(0) + \xi_0(\tau), 0, 0\big)\end{aligned} \tag{5.23}$$

for $\tau \geq 0$ and

$$a_1(0)\xi_0(0) + a_2(0)\eta_0(0) = \alpha(0) - a_1(0)X_0(0). \tag{5.24}$$

Proceeding freely, we have

$$\frac{d\eta_0}{d\tau} = \frac{g_2\big(X_0(0) + \xi_0(\tau), 0, 0\big)}{f_2\big(X_0(0) + \xi_0(\tau), 0, 0\big)} \frac{d\xi_0}{d\tau}$$

and integrating

$$\eta_0(\tau) = \int_0^{\xi_0(\tau)} \frac{g_2\big(X_0(0) + r, 0, 0\big)}{f_2\big(X_0(0) + r, 0, 0\big)}\, dr \tag{5.25}$$

since both ξ_0 and $\eta_0 \to 0$ as $\tau \to \infty$. Thus, the initial condition implies that $\xi_0(0)$ must be a root λ of the nonlinear equation

$$\begin{aligned} Q(\lambda) \equiv a_1(0)\lambda + a_2(0) \int_0^{\lambda} \frac{g_2\big(X_0(0) + r, 0, 0\big)}{f_2\big(X_0(0) + r, 0, 0\big)}\, dr - \alpha(0) \\ + a_1(0) X_0(0) \\ = 0. \end{aligned}$$

Picking some root λ, ξ_0 will satisfy the nonlinear initial value problem

$$\frac{d\xi_0}{d\tau} = f_2\big(X_0(0) + \xi_0, 0, 0\big) \int_0^{\xi_0} \left(\frac{g_2\big(X_0(0) + r, 0, 0\big)}{f_2\big(X_0(0) + r, 0, 0\big)} \right) dr$$

$$\xi_0(0) = \lambda. \tag{5.26}$$

In particular, $\xi_0(\tau) = \eta_0(\tau) = 0$ if $\lambda = 0$. Otherwise, additional hypotheses are necessary. Thus, we will assume

(Hii) *The equation*

$$\int_0^{\lambda} \left[a_1(0) + a_2(0) \frac{g_2\big(X_0(0) + r, 0, 0\big)}{f_2\big(X_0(0) + r, 0, 0\big)} \right] dr = \alpha(0) - a_1(0) X_0(0) \tag{5.27}$$

has a solution λ *such that*

$$a_1(0) + a_2(0) \frac{g_2\big(X_0(0) + \lambda, 0, 0\big)}{f_2\big(X_0(0) + \lambda, 0, 0\big)} \neq 0. \tag{5.28}$$

Further, for some $\kappa > 0$, suppose

$$g_2(R,0,0) \le -\kappa \tag{5.29}$$

and

$$|f_2(R,0,0)| \ge \kappa \tag{5.30}$$

for all values of R between $X_0(0)$ and $X_0(0) + \lambda$.

We note that (5.28) implies that λ is a simple root of (5.27), i.e., $Q'(\lambda) \neq 0$. Further for the problem (5.16) hypothesis (Hii) simply requires that (5.8) hold and there is then only one determination of λ. In general, (5.26), (5.27), and (5.30) imply that $(1/\xi_0)(d\xi_0/d\tau)$ is negative, so $|\xi_0(\tau)|$ decreases monotonically as τ increases. Thus, ξ_0 will exist for $\tau \ge 0$ and we will have

$$|\xi_0(\tau)| \le |\lambda| e^{-\sigma\tau} \qquad \text{for} \quad \tau \ge 0,$$

where σ is such that

$$f_2(R,0,0)\,\frac{g_2(S,0,0)}{f_2(S,0,0)} \le -\sigma < 0$$

for all values R and S between $X_0(0)$ and $X_0(0) + \lambda$. The estimate for ξ_0 implies that $|\eta_0(\tau)|$ also decreases monotonically such that

$$\eta_0(\tau) = O(e^{-\sigma\tau}) \qquad \text{as} \quad \tau \to \infty.$$

In general, of course, it is necessary to solve (5.26) by successive approximations because the differential equation cannot be explicitly integrated. Knowing ξ_0, however, η_0 is given by (5.25).

From higher-order terms in (5.21), we obtain the linear variable coefficient system

$$\begin{aligned}
\frac{d\xi_j}{d\tau} &= \eta_j(\tau) f_2\Big(X_0(0) + \xi_0(\tau), 0, 0\Big) \\
&\quad + \xi_j(\tau) f_{2x}\Big(X_0(0) + \xi_0(\tau), 0, 0\Big)\, \eta_0(\tau) + \tilde{P}_{j-1}(\tau) \\
\frac{d\eta_j}{d\tau} &= \eta_j(\tau)\, g_2\Big(X_0(0) + \xi_0(\tau), 0, 0\Big) \\
&\quad + \xi_j(\tau)\, g_{2x}\Big(X_0(0) + \xi_0(\tau), 0, 0\Big)\, \eta_0(\tau) + \tilde{Q}_{j-1}(\tau),
\end{aligned} \tag{5.31}$$

where $\tilde{P}_{j-1}$ and $\tilde{Q}_{j-1}$ are known successively and satisfy

$$\tilde{P}_{j-1}(\tau) = O(e^{-\tilde{\sigma}\tau}) = \tilde{Q}_{j-1}(\tau) \qquad \text{as} \quad \tau \to \infty$$

for any $\tilde{\sigma}$ such that $0 < \tilde{\sigma} < \min(\sigma, \kappa)$. Further, the initial condition (5.22) implies that

$$a_1(0)\,\xi_j(0) + a_2(0)\,\eta_j(0) = \tilde{\alpha}_{j-1}, \tag{5.32}$$

where $\tilde{\alpha}_{j-1}$ is also known successively. Rearranging (5.31), we have

$$\frac{d\eta_j}{d\tau} = \frac{d}{d\tau}\left(\frac{\xi_j g_2\big(X_0(0) + \xi_0(\tau), 0, 0\big)}{f_2\big(X_0(0) + \xi_0(\tau), 0, 0\big)}\right) + \tilde{\tilde{Q}}_{j-1}(\tau),$$

where $\tilde{\tilde{Q}}_{j-1}$ is known and exponentially decaying. Since both ξ_j and $\eta_j \to 0$ as $\tau \to \infty$,

$$\eta_j(\tau) = \xi_j(\tau)\frac{g_2\big(X_0(0) + \xi_0(\tau), 0, 0\big)}{f_2\big(X_0(0) + \xi_0(\tau), 0, 0\big)} - \int_\tau^\infty \tilde{\tilde{Q}}_{j-1}(s)\,ds \tag{5.33}$$

and, using (5.31) again, ξ_j must satisfy

$$\frac{d}{d\tau}\left(\frac{\xi_j}{f_2\big(X_0(0) + \xi_0(\tau), 0, 0\big)}\right) = \xi_j\frac{g_2\big(X_0(0) + \xi_0(\tau), 0, 0\big)}{f_2\big(X_0(0) + \xi_0(\tau), 0, 0\big)} + \tilde{\tilde{P}}_{j-1}(\tau),$$

where $\tilde{\tilde{P}}_{j-1}$ is known and exponentially decaying. Thus,

$$\xi_j(\tau) = f_2\big(X_0(0) + \xi_0(\tau), 0, 0\big) \times\left[\frac{\xi_j(0)\exp\left[\int_0^\tau g_2(X_0(0) + \xi_0(s), 0, 0)\,ds\right]}{f_2(X_0(0) + \lambda, 0, 0)} + \int_0^\tau \exp\left[\int_r^\tau g_2(X_0(0) + \xi_0(s), 0, 0)\,ds\right]\tilde{\tilde{P}}_{j-1}(r)\,dr\right]. \tag{5.34}$$

Using condition (5.28), then, $\xi_j(0)$ is uniquely determined and so are $\xi_j(\tau)$ and $\eta_j(\tau)$. Finally, since $g_2(X_0(0) + \xi_0(\tau), 0, 0) \leq -\kappa < 0$, it follows that

$$\xi_j(\tau) = O(e^{-\tilde{\sigma}\tau}) = \eta_j(\tau) \qquad \text{as} \quad \tau \to \infty,$$

where $\tilde{\sigma} > 0$. Thus, we have

THEOREM 5: *Under hypotheses* (Hi) *and* (Hii), *the boundary value problem* (5.14)–(5.15) *has a solution* $(x(t,\varepsilon), y(t,\varepsilon))$ *for* ε *sufficiently small which is such that, for each integer* $N \geq 0$,

$$x(t,\varepsilon) = \sum_{j=0}^{N} \left(X_j(t) + \xi_j(t/\varepsilon)\right)\varepsilon^j + \varepsilon^{N+1} R(t,\varepsilon)$$

$$y(t,\varepsilon) = \frac{1}{\varepsilon}\eta_0(t/\varepsilon) + \sum_{j=0}^{N} \left(Y_j(t) + \eta_{j+1}(t/\varepsilon)\right)\varepsilon^j + \varepsilon^{N+1} S(t,\varepsilon),$$

where both $R(t,\varepsilon)$ *and* $S(t,\varepsilon)$ *are uniformly bounded throughout* $0 \leq t \leq 1$.

Remarks

1. The asymptotic solution (x, y) is not uniquely determined because several values $X_0(1)$ may be possible in hypothesis (Hi) and, for each $X_0(1)$, several values λ may satisfy hypothesis (Hii). For example, the problem

$$\frac{dx}{dt} = y$$

$$\varepsilon\frac{dy}{dt} = -\tfrac{1}{2}(1 + 3x^2)y$$

$$x(0,\varepsilon) + \varepsilon y(0,\varepsilon) = 0, \qquad x(1,\varepsilon) = 0$$

has $(X_0(t), Y_0(t)) \equiv (0,0)$ as the unique solution of its reduced problem. (The outer expansion has all terms zero.) However, the corresponding $Q(\lambda) = \int_0^\lambda (1 - \frac{1}{2}(1 + 3r^2))\,dr$ has the three zeros $\lambda = 0$, $\lambda = 1$, and $\lambda = -1$, each of which satisfies hypothesis (Hii). The three corresponding asymptotic solutions are

$$x(t,\varepsilon) = \lambda e^{-t/2\varepsilon}[(1+\lambda^2) - \lambda^2 e^{-t/\varepsilon}]^{-1/2}$$

$$y(t,\varepsilon) = -\frac{\lambda}{2\varepsilon}(1+\lambda^2)e^{-t/2\varepsilon}[(1+\lambda^2) - \lambda^2 e^{-t/\varepsilon}]^{-3/2}.$$

They all converge for $t > 0$ to the trivial solution of the reduced problem. Related boundary value problems with multiple solutions are discussed in Chapter 7, and by Chen (1972) and O'Malley (1972c). We note that problems where $X_0(1)$ is not a simple root of $P(\gamma)$ and/or where $\xi_0(0)$ is not a simple root of $Q(\lambda)$ are also tractable, but the resulting expansions will necessarily be more complicated than (5.20).

2. A proof of asymptotic correctness is given by O'Malley (1970c). In cases where $g_2(x,t,0) > 0$, one can obtain similar results with nonuniform convergence at $t = 1$.

3. Several related problems can be solved analogously. For example, if

$$a_2(\varepsilon) \sim \frac{a_{20}}{\varepsilon} + \sum_{j=0}^{\infty} a_{2,j+1}\varepsilon^j \qquad \text{as} \quad \varepsilon \to 0$$

with $a_{20} \neq 0$, the problem (5.14)–(5.15) can be asymptotically solved using only hypothesis (Hi) [see O'Malley (1970c)]. Likewise, if $f_2(X_0(0),0,0) = f_{2x}(X_0(0),0,0) = 0$ and $a_2(0) \neq 0$, the problem can also be solved asymptotically under hypothesis (Hi) with $\xi_0(\tau) \equiv 0$.

4. Earlier work on boundary value problems for quasi-linear systems includes that of Harris (1962) and Macki (1967). Some general results for vector systems are given in Hoppensteadt (1971).

4. AN EXTENDED DISCUSSION OF A NONLINEAR EXAMPLE

Consider the two-point problem

$$\varepsilon x'' = (x')^2, \qquad x(0) = 0, \quad x(1) = 1 \tag{5.35}$$

whose solution is readily found to be

$$x(t, \varepsilon) = -\varepsilon \ln(1 - t + te^{-1/\varepsilon}). \tag{5.36}$$

Away from $t = 1$, note that $x(t, \varepsilon)$ converges to the limiting solution

$$\tilde{X}(t, \varepsilon) = -\varepsilon \ln(1 - t) \tag{5.37}$$

as the small positive parameter ε tends to zero [and that $\tilde{X}(t, 0) = 0$ satisfies the reduced equation]. This limit, however, cannot approximate the solution near $t = 1$, where it blows up logarithmically. Thus, nonuniform convergence of the solution must occur at $t = 1$ as $\varepsilon \to 0$.

If we did not know the exact solution, how might we attempt to find an approximate one? Since the reduced equation $(x')^2 = 0$ has only constant solutions, no solution of it can be a uniformly valid limiting solution throughout $0 \leq t \leq 1$. Moreover, the results of Vishik and Lyusternik (cf. Section 5.1) indicate that endpoint nonuniformities are possible with derivatives being unbounded there like $O(e^{C/\varepsilon})$ for some $C > 0$. [Indeed, $x'(1, \varepsilon) = \varepsilon e^{1/\varepsilon} - \varepsilon$.] It would be reasonable to also study the possibility of nonuniform convergence within $0 < t < 1$, but we shall not do so. How to predict that the nonuniform convergence occurs at $t = 1$ is not obvious. We note, however, that the differential equation implies that x is nondecreasing. By the boundary conditions, then, $\dot{x}(1)$ must be positive. Noting that the linear equation $\varepsilon\ddot{x} = a\dot{x}$ for $a = \dot{x}(1) > 0$ allows boundary layer behavior at the right endpoint of any interval, it is reasonable to anticipate the limiting behavior obtained. Thus, we shall ask that the solution $x(t, \varepsilon)$ be asymptotically represented in the form

$$x(t, \varepsilon) = X(t, \varepsilon) + \xi(\kappa, \varepsilon), \tag{5.38}$$

where the outer solution $X(t, \varepsilon)$ converges to the limiting solution $-\varepsilon \ln(1 - t)$ as $\varepsilon \to 0$ and $\xi(\kappa, \varepsilon)$, the boundary layer correction at $t = 1$, tends to zero as the stretched variable $\kappa = (1 - t)/\phi(\varepsilon)$, for some $\phi(\varepsilon) = o(1)$, tends to infinity. Vishik and Lyusternik's results indicate that a selection $\phi(\varepsilon) = e^{-C/\varepsilon}$, $C > 0$, would be appropriate since $x'(1,\varepsilon) \sim \xi(0,\varepsilon)/\phi(\varepsilon)$. Thus, we will proceed with

$$\kappa = \frac{1-t}{e^{-1/\varepsilon}}, \tag{5.39}$$

noting that the usual technique involving use of principal limits does not seem to provide the necessary stretched variable in this example.

A new feature here is that the outer solution blows up near the point $t = 1$ where the boundary layer occurs. Since $x(1) = 1$, however, the boundary layer correction $\xi(\kappa,\varepsilon)$ must also blow up at $\kappa = 0$ in such a way that the singularity of the outer solution is canceled. Similar behavior is known to occur in certain physical problems [cf., e.g., Dickey (1973) for a problem involving the inflated toroidal membrane].

Clearly, the outer solution $X(t,\varepsilon)$ must satisfy the system for $t < 1$ and the initial condition. Thus, we have

$$\varepsilon X_{tt} - X_t^2 = 0, \qquad X(0,\varepsilon) = 0. \tag{5.40}$$

When $\varepsilon = 0$, $X_t(t,0) = 0 = X(0,0)$, so $X(t,0) = 0$. Thus, it is natural to set

$$X(t,\varepsilon) = \varepsilon Z(t,\varepsilon),$$

where

$$Z_{tt} = Z_t^2, \qquad Z(0,\varepsilon) = 0.$$

(Solving the differential equation for Z is no simpler than solving the original equation for x, but our point here is to illustrate that our asymptotic representation—outer solution plus correction—still holds.) Integrating, then, we have the outer solution

$$X(t,\varepsilon) = \varepsilon Z(t,\varepsilon) = -\varepsilon \ln\left(\frac{c(\varepsilon)-t}{c(\varepsilon)}\right), \tag{5.41}$$

where $c(\varepsilon)$ is an undetermined smooth function of ε. For $Z(t,\varepsilon)$ to be defined throughout $0 \le t < 1$, we require that $c(\varepsilon) \ge 1$. [Since the outer solution is asymptotically given by (5.41) while $x(1) = 1$ is prescribed, we have another indication that an exponential stretching near $t = 1$ is necessary.]

Having obtained the outer solution [up to $c(\varepsilon)$], we shall now seek the boundary layer correction. Note that the differential equation [cf. (5.35)] implies that

$$\varepsilon X_{tt} + \varepsilon e^{2/\varepsilon}\xi_{\kappa\kappa} = X_t^2 - 2e^{1/\varepsilon}X_t\xi_\kappa + e^{2/\varepsilon}\xi_\kappa^2$$

or

$$\xi_\kappa^2 = \varepsilon\left(\xi_{\kappa\kappa} + \frac{2e^{-1/\varepsilon}\xi_\kappa c(\varepsilon)}{c(\varepsilon) - t}\right). \tag{5.42}$$

When $\varepsilon = 0$, then, $\xi_\kappa(\kappa, 0) = 0$ and, since $\xi \to 0$ as $\kappa \to \infty$, $\xi(\kappa, 0) = 0$. Thus, we set

$$\xi(\kappa, \varepsilon) = \varepsilon\eta(\kappa, \varepsilon),$$

where

$$\eta_\kappa^2 = \eta_{\kappa\kappa} + \left[\frac{2c(\varepsilon)}{\kappa + \big(c(\varepsilon) - 1\big)e^{1/\varepsilon}}\right]\eta_\kappa.$$

Integrating, then, $v = 1/\eta_\kappa$ satisfies

$$v_\kappa - \left(\frac{2c(\varepsilon)}{\kappa + \big(c(\varepsilon) - 1\big)e^{1/\varepsilon}}\right)v = -1.$$

Thus, if $(c(\varepsilon) - 1)e^{1/\varepsilon} \to \infty$ as $\varepsilon \to 0$, the limiting equation is $v_\kappa(\kappa, 0) = -1$ and $\eta(\kappa, 0) = l_1 - \ln(\kappa - l_2)$ for constants l_1 and l_2. This is unsatisfactory, however, since η must remain bounded as $\kappa \to \infty$. Thus, we take

$$c(\varepsilon) = 1 + o(e^{-1/\varepsilon}) \tag{5.43}$$

for which the limiting system $v_\kappa(\kappa, 0) - (2/\kappa)v(\kappa, 0) = -1$ implies

$$\eta(\kappa, 0) = l_1 + \ln\left(\frac{\kappa}{\kappa + l_2}\right), \tag{5.44}$$

where l_1 and l_2 are undetermined constants. Since $\eta \to 0$ as $\kappa \to \infty$, however, $l_1 = 0$. So far, then, we have

$$x(t, \varepsilon) = -\varepsilon \ln(1 - t) + \varepsilon \ln \kappa - \varepsilon \ln(\kappa + l_2) + o(\varepsilon)$$

and, expressing t in terms of κ, the singularities at $t = 1$ and $\kappa = 0$ cancel and we have

$$x(t, \varepsilon) = 1 - \varepsilon \ln(\kappa + l_2) + o(\varepsilon).$$

To satisfy the boundary condition at $t = 1 (\kappa = 0)$, however, we must pick $l_2 = 1$. Thus, we have

$$x(t, \varepsilon) = -\varepsilon \ln(1 - t) + \varepsilon \ln\left(\frac{\kappa}{1 + \kappa}\right) + o(\varepsilon) \tag{5.45}$$

uniformly in $0 \leq t \leq 1$, in agreement with the known solution (5.36). Moreover, the outer solution, which is asymptotically valid for $t < 1$, is given by $X(t, \varepsilon) = -\varepsilon \ln(1 - t)$.

CHAPTER 6

THE SINGULARLY PERTURBED LINEAR STATE REGULATOR PROBLEM

We shall consider the linear state regulator problem consisting of the system

$$\begin{aligned} \frac{dx}{dt} &= A_1(t,\varepsilon)x + A_2(t,\varepsilon)z + B_1(t,\varepsilon)u \\ \varepsilon\frac{dz}{dt} &= A_3(t,\varepsilon)x + A_4(t,\varepsilon)z + B_2(t,\varepsilon)u \end{aligned} \tag{6.1}$$

on the interval $0 \leq t \leq 1$ (or, equivalently, on any closed bounded interval), the initial conditions

$$\begin{aligned} x(0,\varepsilon) &= x^0(\varepsilon) \\ z(0,\varepsilon) &= z^0(\varepsilon), \end{aligned} \tag{6.2}$$

and the quadratic cost functional

$$J(\varepsilon) = \tfrac{1}{2}y'(1,\varepsilon)\,\pi(\varepsilon)\,y(1,\varepsilon) + \tfrac{1}{2}\int_0^1 [y'(t,\varepsilon)\,Q(t,\varepsilon)\,y(t,\varepsilon) + u'(t,\varepsilon)\,R(t,\varepsilon)\,u(t,\varepsilon)]\,dt \tag{6.3}$$

which is to be minimized. Here, the prime denotes transposition,

$$y = \begin{bmatrix} x \\ z \end{bmatrix}, \quad Q = \begin{bmatrix} Q_1 & Q_2 \\ Q_2' & Q_3 \end{bmatrix}, \quad \text{and} \quad \pi = \begin{bmatrix} \pi_1 & \varepsilon\pi_2 \\ \varepsilon\pi_2' & \varepsilon\pi_3 \end{bmatrix},$$

where x is an n-vector, z is an m-vector, u is an r-vector, and the matrix R is symmetric and positive definite while the matrices Q and π are symmetric and positive semidefinite throughout $0 \leq t \leq 1$ for $\varepsilon \geq 0$ small. Crudely, the object is to find a control u which is not too large which will drive the state y toward zero, especially at the terminal time $t = 1$.

When $\varepsilon > 0$ is fixed, this elementary control problem has a unique optimal solution which minimizes J [cf., e.g., Athans and Falb (1966) or Anderson and Moore (1971) for appropriate background material]. We are interested, however, in obtaining the asymptotic solution of the problem as the small positive parameter ε tends to zero. Such singular perturbation problems are of considerable importance in practical situations where ε represents certain often neglected "parasitic" parameters whose presence causes the order of the mathematical model to increase [cf. Sannuti and Kokotović (1969) and Kokotović (1972)]. In particular, we note that Hadlock *et al.* (1970) give an example for the optimal tension regulation of a strip winding process where asymptotic results are far superior to the physically unacceptable results obtained by setting $\varepsilon = 0$.

We will assume that the matrices A_i, B_i, x^0, z^0, π_i, Q_i, and R all have asymptotic power series expansions as $\varepsilon \to 0$ and that the coefficients in the expansions for the A_i, B_i, Q_i, and R are infinitely differentiable functions for $0 \leq t \leq 1$.

To obtain necessary and sufficient conditions for an optimal control, we introduce the Hamiltonian

$$H(x, z, u, p_1, p_2, t, \varepsilon)$$
$$= \tfrac{1}{2}(x'Q_1 x + 2x'Q_2 z + z'Q_3 z + u'Ru)$$
$$+ p_1'(A_1 x + A_2 z + B_1 u) + p_2'(A_3 x + A_4 z + B_2 u). \quad (6.4)$$

Elementary calculus of variations [cf. Kalman *et al.* (1969)] implies that along an optimal trajectory

$$\frac{\partial H}{\partial u} = Ru + B_1' p_1 + B_2' p_2 = 0, \quad (6.5)$$

where the costates p_1 and εp_2 are defined to be solutions of the differential equations

$$\begin{aligned} \frac{dp_1}{dt} &= -\frac{\partial H}{\partial x} = -Q_1 x - Q_2 z - A_1' p_1 - A_3' p_2 \\ \varepsilon \frac{dp_2}{dt} &= -\frac{\partial H}{\partial z} = -Q_2' x - Q_3 z - A_2' p_1 - A_4' p_2 \end{aligned} \quad (6.6)$$

on $0 \leq t \leq 1$ and the terminal conditions

$$\begin{pmatrix} p_1(1,\varepsilon) \\ \varepsilon p_2(1,\varepsilon) \end{pmatrix} = \pi(\varepsilon) \begin{pmatrix} x(1,\varepsilon) \\ z(1,\varepsilon) \end{pmatrix};$$

i.e.,

$$\begin{aligned} p_1(1,\varepsilon) &= \pi_1(\varepsilon) x(1,\varepsilon) + \varepsilon \pi_2(\varepsilon) z(1,\varepsilon) \\ p_2(1,\varepsilon) &= \pi_2'(\varepsilon) x(1,\varepsilon) + \pi_3(\varepsilon) z(1,\varepsilon). \end{aligned} \quad (6.7)$$

We note that the state equations (6.1) can be rewritten as

$$\frac{dx}{dt} = \frac{\partial H}{\partial p_1}, \qquad \varepsilon \frac{dz}{dt} = \frac{\partial H}{\partial p_2},$$

analogous to the usual Hamilton–Jacobi theory [cf. Courant and Hilbert (1962)]. Further (6.5) implies the control relation

$$u(t,\varepsilon) = -R^{-1}(t,\varepsilon)\Big(B_1'(t,\varepsilon) p_1(t,\varepsilon) + B_2'(t,\varepsilon) p_2(t,\varepsilon)\Big) \quad (6.8)$$

and since $\partial^2 H/\partial u^2 = R$ is positive definite, this optimal control will minimize $J(\varepsilon)$.

Using (6.8), the $m + n$ state equations (6.1) and the $m + n$ costate equations (6.6) can be rewritten as the linear system

$$\frac{dx}{dt} = A_1 x + A_2 z - S_1 p_1 - S p_2 \tag{6.9a}$$

$$\frac{dp_1}{dt} = -Q_1 x - Q_2 z - A_1' p_1 - A_3' p_2 \tag{6.9b}$$

$$\varepsilon \frac{dz}{dt} = A_3 x + A_4 z - S' p_1 - S_2 p_2 \tag{6.9c}$$

$$\varepsilon \frac{dp_2}{dt} = -Q_2' x - Q_3 z - A_2' p_1 - A_4' p_2, \tag{6.9d}$$

where

$$S(t, \varepsilon) = B_1(t, \varepsilon) R^{-1}(t, \varepsilon) B_2'(t, \varepsilon)$$

and

$$S_i(t, \varepsilon) = B_i(t, \varepsilon) R^{-1}(t, \varepsilon) B_i'(t, \varepsilon), \qquad i = 1, 2.$$

We must solve the system (6.9) subject to the $2m + 2n$ boundary conditions provided by the initial conditions (6.2) and the terminal conditions (6.7).

Since the order of the system drops from $2m + 2n$ for $\varepsilon > 0$ to $2n$ for $\varepsilon = 0$, we have a singularly perturbed problem. In particular, the system obtained when $\varepsilon = 0$ cannot be expected, in general, to satisfy all the limiting boundary conditions. Thus, even if the limiting solution as $\varepsilon \to 0$ satisfies the limiting system, nonuniform convergence near the endpoints must be expected as $\varepsilon \to 0$. From our previous work, we are led to seek an asymptotic solution of the form

$$x(t, \varepsilon) = X(t, \varepsilon) + \varepsilon m_1(\kappa, \varepsilon) + \varepsilon n_1(\sigma, \varepsilon) \tag{6.10a}$$

$$z(t, \varepsilon) = Z(t, \varepsilon) + m_2(\kappa, \varepsilon) + n_2(\sigma, \varepsilon) \tag{6.10b}$$

$$p_1(t, \varepsilon) = P_1(t, \varepsilon) + \varepsilon \rho_1(\kappa, \varepsilon) + \varepsilon \gamma_1(\sigma, \varepsilon) \tag{6.10c}$$

$$p_2(t, \varepsilon) = P_2(t, \varepsilon) + \rho_2(\kappa, \varepsilon) + \gamma_2(\sigma, \varepsilon), \tag{6.10d}$$

where κ and σ are the stretched variables

$$\kappa = t/\varepsilon, \qquad \sigma = \frac{1 - t}{\varepsilon}; \tag{6.11}$$

$$\big(X(t, \varepsilon), Z(t, \varepsilon), P_1(t, \varepsilon), P_2(t, \varepsilon)\big) \tag{6.12}$$

will be the outer solution;

$$\big(\varepsilon m_1(\kappa,\varepsilon), m_2(\kappa,\varepsilon), \varepsilon\rho_1(\kappa,\varepsilon), \rho_2(\kappa,\varepsilon)\big) \tag{6.13}$$

will be the boundary layer correction at $t = 0$; and

$$\big(\varepsilon n_1(\sigma,\varepsilon), n_2(\sigma,\varepsilon), \varepsilon\gamma_1(\sigma,\varepsilon), \gamma_2(\sigma,\varepsilon)\big) \tag{6.14}$$

will be the boundary layer correction at $t = 1$. The three matrices (6.12)–(6.14) will all have asymptotic power series expansions. Further, the terms in the expansions for the boundary layer corrections will tend to zero as the appropriate stretched variable tends to infinity. Away from $t = 0$ and $t = 1$, then, the solution will be asymptotically given by the outer solution, while at these endpoints convergence will be nonuniform as $\varepsilon \to 0$.

Clearly, then, we must ask that the outer solution satisfy the linear system (6.9). Setting $\varepsilon = 0$, the leading term of the outer solution must satisfy the reduced system

$$\frac{dX_0}{dt} = A_{10}X_0 + A_{20}Z_0 - S_{10}P_{10} - S_0P_{20} \tag{6.15a}$$

$$\frac{dP_{10}}{dt} = -Q_{10}X_0 - Q_{20}Z_0 - A'_{10}P_{10} - A'_{30}P_{20} \tag{6.15b}$$

$$0 = A_{30}X_0 + A_{40}Z_0 - S'_0P_{10} - S_{20}P_{20} \tag{6.15c}$$

$$0 = -Q'_{20}X_0 - Q_{30}Z_0 - A'_{20}P_{10} - A'_{40}P_{20}. \tag{6.15d}$$

Applying two of the boundary conditions when $\varepsilon = 0$ further requires that

$$X_0(0) = x_0^0$$
$$P_{10}(1) = \pi_{10}X_0(1). \tag{6.16}$$

To solve (6.15)–(6.16), it is necessary to solve Eq. (6.15c,d) for Z_0 and P_{20} as linear functions of X_0 and P_{10} and to solve the remaining boundary value problem for X_0 and P_{10}. This will require two matrices to be nonsingular. As our first assumption, then, we will ask:

(Hi) *The reduced problem*

$$\frac{dx}{dt} = A_1(t,0)x + A_2(t,0)z - B_1(t,0)R^{-1}(t,0)B_1'(t,0)p_1 - B_1(t,0)R^{-1}(t,0)B_2'(t,0)p_2 \tag{6.17a}$$

$$\frac{dp_1}{dt} = Q_1(t,0)x - Q_2(t,0)z - A_1'(t,0)p_1 - A_3'(t,0)p_2 \tag{6.17b}$$

$$0 = A_3(t,0)x + A_4(t,0)z - B_2(t,0)R^{-1}(t,0)B_1'(t,0)p_1 - B_2(t,0)R^{-1}(t,0)B_2'(t,0)p_2 \tag{6.17c}$$

$$0 = -Q_2'(t,0)x - Q_3(t,0)z - A_2'(t,0)p_1 - A_4'(t,0)p_2 \tag{6.17d}$$

$$x(0) = x_0^0, \qquad p_1(1) - \pi_1(0)x(1) = 0 \tag{6.17e}$$

has a unique solution

$$\big(X_0(t), Z_0(t), P_{10}(t), P_{20}(t)\big)$$

throughout $0 < t < 1$.

In particular, then,

$$\begin{pmatrix} Z_0 \\ P_{20} \end{pmatrix} = \begin{pmatrix} A_{40} & -S_{20} \\ -Q_{30} & -A_{40}' \end{pmatrix}^{-1} \begin{pmatrix} -A_{30} & S_0' \\ Q_{20}' & A_{20}' \end{pmatrix} \begin{pmatrix} X_0 \\ P_{10} \end{pmatrix}$$

and there will remain the linear system

$$\begin{pmatrix} \dot{X}_0 \\ \dot{P}_{10} \end{pmatrix} = \left[\begin{pmatrix} A_{10} & -S_{10} \\ -Q_{10} & -A_{10}' \end{pmatrix} + \begin{pmatrix} A_{20} & -S_0 \\ -Q_{20} & -A_{30}' \end{pmatrix} \times \begin{pmatrix} A_{40} & -S_{20} \\ -Q_{30} & -A_{40}' \end{pmatrix}^{-1} \begin{pmatrix} -A_{30} & S_0' \\ Q_{20}' & A_{20}' \end{pmatrix} \right] \begin{pmatrix} X_0 \\ P_{10} \end{pmatrix}$$

subject to the boundary conditions on X_0 and P_{10}. Note that, since $Z_0(0) \neq z_0^0$ and $P_{20}(1) \neq \pi_{20}' X_0(1) + \pi_{30} Z_0(1)$, in general, nonuniform convergence generally occurs at both $t = 0$ and $t = 1$.

Higher-order terms $(X_j(t), Z_j(t), P_{1j}(t), P_{2j}(t))$ of the outer expansion will satisfy nonhomogeneous forms of (6.15)–(6.16) which can be uniquely solved successively up to specification of $X_j(0)$ and $P_{1j}(1)$

$- \pi_{10} X_j(1)$. Since

$$X(0,\varepsilon) = x^0(\varepsilon) - \varepsilon m_1(0,\varepsilon)$$

and

$$P_1(1,\varepsilon) - \pi_1(\varepsilon)X(1,\varepsilon)$$
$$= \varepsilon\Big(-\gamma_1(0,\varepsilon) + \pi_1(\varepsilon)n_1(0,\varepsilon) + \pi_2(\varepsilon)(Z(1,\varepsilon) + n_2(0,\varepsilon))\Big),$$

the terms of the outer expansion can be determined successively once lower-order terms in the initial values of the boundary layer corrections $m_1(0,\varepsilon)$, $\gamma_1(0,\varepsilon)$, $n_1(0,\varepsilon)$, and $n_2(0,\varepsilon)$ are known.

Since the boundary layer correction at $t = 1$ (or 0) is asymptotically negligible at $t = 0$ (or 1), the boundary layer correction at $t = 0$ must satisfy the linear system

$$\frac{dm_1}{d\kappa} = \varepsilon A_1(\varepsilon\kappa,\varepsilon)m_1 + A_2(\varepsilon\kappa,\varepsilon)m_2 - \varepsilon S_1(\varepsilon\kappa,\varepsilon)\rho_1 - S(\varepsilon\kappa,\varepsilon)\rho_2 \tag{6.18a}$$

$$\frac{d\rho_1}{d\kappa} = -\varepsilon Q_1(\varepsilon\kappa,\varepsilon)m_1 - Q_2(\varepsilon\kappa,\varepsilon)m_2 - \varepsilon A_1'(\varepsilon\kappa,\varepsilon)\rho_1 - A_3'(\varepsilon\kappa,\varepsilon)\rho_2 \tag{6.18b}$$

$$\frac{dm_2}{d\kappa} = \varepsilon A_3(\varepsilon\kappa,\varepsilon)m_1 + A_4(\varepsilon\kappa,\varepsilon)m_2 - \varepsilon S'(\varepsilon\kappa,\varepsilon)\rho_1 - S_2(\varepsilon\kappa,\varepsilon)\rho_2 \tag{6.18c}$$

$$\frac{d\rho_2}{d\kappa} = -\varepsilon Q_2'(\varepsilon\kappa,\varepsilon)m_1 - Q_3(\varepsilon\kappa,\varepsilon)m_2 - \varepsilon A_2'(\varepsilon\kappa,\varepsilon)\rho_1 - A_4'(\varepsilon\kappa,\varepsilon)\rho_2 \tag{6.18d}$$

while the boundary layer correction at $t = 1$ must satisfy

$$\frac{dn_1}{d\sigma} = -\varepsilon A_1(1-\varepsilon\sigma,\varepsilon)n_1 - A_2(1-\varepsilon\sigma,\varepsilon)n_2 + \varepsilon S_1(1-\varepsilon\sigma,\varepsilon)\gamma_1 + S(1-\varepsilon\sigma,\varepsilon)\gamma_2 \tag{6.19a}$$

$$\frac{d\gamma_1}{d\sigma} = \varepsilon Q_1(1-\varepsilon\sigma,\varepsilon)n_1 + Q_2(1-\varepsilon\sigma,\varepsilon)n_2 + \varepsilon A_1'(1-\varepsilon\sigma,\varepsilon)\gamma_1 + A_3'(1-\varepsilon\sigma,\varepsilon)\gamma_2 \tag{6.19b}$$

$$\frac{dn_2}{d\sigma} = -\varepsilon A_3(1-\varepsilon\sigma,\varepsilon)n_1 - A_4(1-\varepsilon\sigma,\varepsilon)n_2 + \varepsilon S'(1-\varepsilon\sigma,\varepsilon)\gamma_1 + S_2(1-\varepsilon\sigma,\varepsilon)\gamma_2 \tag{6.19c}$$

$$\frac{d\gamma_2}{d\sigma} = \varepsilon Q_2'(1-\varepsilon\sigma,\varepsilon)n_1 + Q_3(1-\varepsilon\sigma,\varepsilon)n_2 + \varepsilon A_2'(1-\varepsilon\sigma,\varepsilon)\gamma_1 + A_4'(1-\varepsilon\sigma,\varepsilon)\gamma_2. \tag{6.19d}$$

When $\varepsilon = 0$, then, we ask that

$$\frac{dm_{10}}{d\kappa} = A_{20}(0)m_{20} - S_0(0)\rho_{20} \tag{6.20a}$$

$$\frac{d\rho_{10}}{d\kappa} = -Q_{20}(0)m_{20} - A'_{30}(0)\rho_{20} \tag{6.20b}$$

$$\frac{dm_{20}}{d\kappa} = A_{40}(0)m_{20} - S_{20}(0)\rho_{20} \tag{6.20c}$$

$$\frac{d\rho_{20}}{d\kappa} = -Q_{30}(0)m_{20} - A'_{40}(0)\rho_{20}. \tag{6.20d}$$

In particular, note that the last two equations of (6.20) form a system of $2m$ linear equations and that the initial condition (6.2) yields the m-vector

$$m_{20}(0) = z_0^0 - Z_0(0). \tag{6.21}$$

In order that m_{20} and $\rho_{20} \to 0$ as $\kappa \to \infty$, it is necessary that the matrix

$$G(t) = \begin{bmatrix} A_{40}(t) & -S_{20}(t) \\ -Q_{30}(t) & -A'_{40}(t) \end{bmatrix}$$

have m eigenvalues with negative real parts at $t = 0$. Analogously, in order to define appropriate determinations of $n_{20}(\sigma)$ and $\gamma_{20}(\sigma)$, it is necessary that $G(1)$ have m eigenvalues with positive real parts. In order to obtain the expansions (6.10), we will use the three additional assumptions:

(H2) *The $2m \times 2m$ matrix*

$$G(t) = \begin{bmatrix} A_4(t,0) & -B_2(t,0)\,R^{-1}(t,0)\,B_2'(t,0) \\ -Q_3(t,0) & -A_4'(t,0) \end{bmatrix} \tag{6.22}$$

has m eigenvalues $\lambda_1(t), \ldots, \lambda_m(t)$ with negative real parts throughout $0 \le t \le 1$ and m eigenvalues $\lambda_{m+1}(t), \ldots, \lambda_{2m}(t)$ with positive real parts there.

(H3) *The linear system*

$$\varepsilon \frac{dc}{dt} = G(t)c \tag{6.23}$$

has m linearly independent solutions of the form

$$c_j(t, \varepsilon) = D_j(t, \varepsilon)e^{\lambda_j(0)t/\varepsilon}, \qquad j = 1, 2, \ldots, m, \tag{6.24}$$

and m linearly independent solutions

$$c_j(t, \varepsilon) = D_j(1 - t, \varepsilon)e^{\lambda_j(1)(t-1)/\varepsilon}, \qquad j = m + 1, \ldots, 2m. \tag{6.25}$$

(H4) *Setting*

$$D_j(0, 0) = \begin{pmatrix} D_j^1 \\ D_j^2 \end{pmatrix}$$

for each j where the D_j^i 's are m-dimensional vectors, suppose that the $m \times m$ matrices

$$D^1 = (D_1^1 \, D_2^1 \cdots D_m^1) \tag{6.26}$$

and

$$D^2 = \big((D_{m+1}^2 - \pi_{30} D_{m+1}^1) \cdots (D_{2m}^2 - \pi_{30} D_{2m}^1)\big) \tag{6.27}$$

are both nonsingular.

Remarks

1. Because of the special form of the matrix $G(t)$, the existence of m eigenvalues λ with negative real parts actually implies the existence of m other eigenvalues $-\lambda$ with positive real parts. Further, when $Q_3 = 0$, (H2) follows provided the real parts of the eigenvalues of $A_4(t, 0)$ are all nonzero throughout $0 \leq t \leq 1$.

2. Hypothesis (H3) follows readily if Eq. (6.23) has no turning points in [0, 1] [cf. Turrittin (1952) and Wasow (1965)]. The construc-

tion of the asymptotic solutions is analogous to that in Section 3.2. Weaker hypotheses are also possible [cf. O'Malley (1972d, 1974) and Kung (1973)].

Since (6.23) can be rewritten $dc/d\kappa = G(\varepsilon\kappa)c$, the linear system for m_{20} and ρ_{20} [cf. (6.20)] has the solution

$$\begin{pmatrix} m_{20}(\kappa) \\ \rho_{20}(\kappa) \end{pmatrix} = \sum_{j=1}^{2m} \beta_j^0 D_j(0,0) e^{\lambda_j(0)\kappa} \tag{6.28}$$

for constants β_j^0, but, since $\lambda_{m+1}, \ldots, \lambda_{2m}$ have positive real parts, we must pick

$$\beta_{m+1}^0 = \beta_{m+2}^0 = \cdots = \beta_{2m}^0 = 0$$

in order to eliminate exponentially growing terms. Further, with $m_{20}(0)$ specified by (6.21), $\beta_1^0, \beta_2^0, \ldots, \beta_m^0$ are uniquely determined since, by (6.26), the matrix of coefficients D^1 is nonsingular. Thus, $\rho_{20}(\kappa)$ and, by (6.20), $dm_{10}/d\kappa$ and $d\rho_{10}/d\kappa$ are exponentially decaying as $\kappa \to \infty$. Defining

$$\begin{aligned} m_{10}(\kappa) &= -\int_\kappa^\infty \frac{dm_{10}}{d\kappa}(s)\,ds \\ \rho_{10}(\kappa) &= -\int_\kappa^\infty \frac{d\rho_{10}}{d\kappa}(s)\,ds \end{aligned} \tag{6.29}$$

the lowest-order terms in the boundary layer correction at $t = 0$ become completely specified as exponentially decaying terms. Analogously, we have

$$\begin{pmatrix} n_{20}(\sigma) \\ \gamma_{20}(\sigma) \end{pmatrix} = \sum_{j=m+1}^{2m} \alpha_j^0 D_j(0,0) e^{-\lambda_j(1)\sigma}, \tag{6.30}$$

where the α_j^0's are uniquely determined by the initial condition

$$\gamma_{20}(0) - \pi_{30} n_{20}(0) = \pi'_{20} X_0(1) + \pi_{30} Z_0(1) \tag{6.31}$$

since the matrix D^2 is nonsingular. It follows that $n_{10}(\sigma)$ and $\gamma_{10}(\sigma)$ are also uniquely determined as exponentially decaying terms.

Higher-order coefficients in the boundary layer corrections are determined as solutions of analogous nonhomogeneous problems. For example, from (6.18), we find that m_{2j} and ρ_{2j} satisfy a system

$$\begin{aligned}\frac{dm_{2j}}{d\kappa} &= A_{40}(0)m_{2j} - S_{20}(0)\rho_{2j} + \mathcal{M}_{j-1}(\kappa)\\ \frac{d\rho_{2j}}{d\kappa} &= -Q_{30}(0)m_{2j} - A'_{40}(0)\rho_{2j} + \mathcal{P}_{j-1}(\kappa),\end{aligned} \tag{6.32}$$

where $\mathcal{M}_{j-1}$ and $\mathcal{P}_{j-1}$ are exponentially decaying and successively known and

$$m_{2j}(0) = z_j^0 - Z_j(0). \tag{6.33}$$

Note that $Z_j(0)$ is known successively since the jth term of the outer expansion is determined by preceding terms of the boundary layer corrections. Since m_{2j} and ρ_{2j} tend to zero as $\kappa \to \infty$, they must be of the form

$$\begin{pmatrix} m_{2j}(\kappa) \\ \rho_{2j}(\kappa) \end{pmatrix} = \sum_{i=1}^{m} \beta_i^j D_i(0,0)\, e^{\lambda_i(0)\kappa} + \begin{pmatrix} \tilde{M}_{2j}(\kappa) \\ \tilde{P}_{2j}(\kappa) \end{pmatrix}, \tag{6.34}$$

where

$$\begin{pmatrix} \tilde{M}_{2j}(\kappa) \\ \tilde{P}_{2j}(\kappa) \end{pmatrix} = \sum_{i=1}^{m} D_i(0,0)\, e^{\lambda_i(0)\kappa} \int_0^{\kappa} \frac{\Delta_i(s)}{\Delta(s)}\, ds + \sum_{i=m+1}^{2m} D_i(0,0)\, e^{\lambda_i(0)\kappa} \int_{\infty}^{\kappa} \frac{\Delta_i(s)}{\Delta(s)}\, ds,$$

$\Delta(s)$ is the determinant

$$\Delta(s) = \det\left(D_1(0,0)\, e^{\lambda_1(0)\kappa} \cdots D_{2m}(0,0)\, e^{\lambda_{2m}(0)\kappa} \right),$$

and $\Delta_i(s)$ is the determinant obtained from $\Delta(s)$ by replacing its ith column by the exponentially decaying vector

$$\begin{pmatrix} \mathcal{M}_{j-1}(s) \\ \mathcal{P}_{j-1}(s) \end{pmatrix}.$$

Since the matrix D^1 is nonsingular the initial condition (6.33) determines (6.34) uniquely. Furthermore, hypothesis (H2) and the fact that $e^{\lambda_i(s)}\Delta_i(s)/\Delta(s)$ is exponentially decaying imply that m_{2j} and ρ_{2j} will also decay exponentially as $\kappa \to \infty$. Integrating the differential equations for $dm_{1j}/d\kappa$ and $d\rho_{1j}/d\kappa$ from κ to infinity, then, we determine the exponentially decaying terms m_{1j} and ρ_{1j}. Proceeding, we can also determine the jth terms in the boundary layer correction at $t = 1$.

Knowing the expansions (6.10) for the costate vectors, the control relation (6.8) implies that the optimal control has an expansion of the form

$$u(t, \varepsilon) = U(t, \varepsilon) + v(\kappa, \varepsilon) + w(\sigma, \varepsilon), \tag{6.35}$$

where U, v, and w have asymptotic power series expansions as $\varepsilon \to 0$ and

$$U_0(t) = -R_0^{-1}(t)[B'_{10}(t)P_{10}(t) + B'_{20}(t)P_{20}(t)]$$

is the control corresponding to the solution of the reduced problem (6.17). Knowing the asymptotic expansions for the optimal trajectories implies that the corresponding power

$$y'(t, \varepsilon)\,Q(t, \varepsilon)y(t, \varepsilon) + u'(t, \varepsilon)\,R(t, \varepsilon)\,u(t, \varepsilon)$$

is of the form

$$L_1(t, \varepsilon) + L_2(\kappa, \varepsilon) + L_3(\sigma, \varepsilon),$$

where the L_i's have asymptotic power series expansions such that L_2(or L_3) tends to zero as κ (or σ) tends to infinity and with

$$\begin{aligned} L_{10}(t) = {} & X'_0(t)\,Q_{10}(t)\,X_0(t) + 2X'_0(t)\,Q_{20}(t)\,Z_0(t) \\ & + Z'_0(t)\,Q_{30}(t)\,Z_0(t) + U'_0(t)\,R_0(t)\,U_0(t). \end{aligned}$$

Thus, the optimal cost can be written as

$$\begin{aligned} J^*(\varepsilon) = {} & \tfrac{1}{2}\lambda(\varepsilon) + \tfrac{1}{2}\int_0^1 L_1(t, \varepsilon)\,dt \\ & + \frac{\varepsilon}{2}\int_0^\infty L_2(\kappa, \varepsilon)\,d\kappa + \frac{\varepsilon}{2}\int_0^\infty L_3(\sigma, \varepsilon)\,d\sigma \end{aligned} \tag{6.36}$$

and it will have an asymptotic expansion

$$J^*(\varepsilon) \sim \sum_{i=0}^{\infty} J_i^* \varepsilon^i$$

with leading term

$$J_0^* = X_0'(1)\pi_{10}X_0(1) + \tfrac{1}{2}\int_0^1 L_{10}(t)\,dt$$

being the cost corresponding to the solution of the reduced problem. Boundary layer corrections, then, contribute leading order terms to the state vector z and to the optimal control u, but their influence on the state vector x and the optimal cost $J^*(\varepsilon)$ is in higher-order terms.

Thus, we have formally obtained the following theorem whose proof is given by O'Malley (1972a).

THEOREM 6. *Under hypotheses* (Hl)–(H4), *the optimal control problem* (6.1)–(6.2) *has a unique solution for* ε *sufficiently small, such that, for every integer* $N \geq 0$, *the optimal control* $u(t, \varepsilon)$ *and the corresponding trajectories* $x(t, \varepsilon)$ *and* $z(t, \varepsilon)$ *satisfy*

$$u(t, \varepsilon) = \sum_{j=0}^{N} \Big(U_j(t) + v_j(\kappa) + w_j(\sigma)\Big)\varepsilon^j + O(\varepsilon^{N+1})$$

$$x(t, \varepsilon) = X_0(t) + \sum_{j=1}^{N} \Big(X_j(t) + m_{1,j-1}(\kappa) + n_{1,j-1}(\sigma)\Big)\varepsilon^j + O(\varepsilon^{N+1})$$

$$z(t, \varepsilon) = \sum_{j=0}^{N} \Big(Z_j(t) + m_{2j}(\kappa) + n_{2j}(\sigma)\Big)\varepsilon^j + O(\varepsilon^{N+1})$$

as $\varepsilon \to 0$ *uniformly in* $0 \leq t \leq 1$. *Here the terms which are functions of* $\kappa = t/\varepsilon$ [*or* $\sigma = (1-t)/\varepsilon$] *decay to zero exponentially as* κ (*or* σ) *tends to infinity* [*i.e., away from* $t = 0$ (*or* 1)]. *Moreover, the optimal cost has an asymptotic expansion such that*

$$J^*(\varepsilon) = \sum_{i=0}^{N} J_i^* \varepsilon^i + O(\varepsilon^{N+1})$$

as $\varepsilon \to 0$.

As an example, consider the scalar problem

$$\frac{dx}{dt} = x$$

$$\varepsilon\frac{dz}{dt} = -x - z + u$$

on the interval $0 \leq t \leq 1$ with $x(0) = x_0$ and $z(0) = z_0$ being prescribed constants and with the quadratic cost

$$J(\varepsilon) = \tfrac{1}{2}\left(x^2(1) + \varepsilon z^2(1)\right) + \tfrac{1}{2}\int_0^1 \left(x^2(t) + u^2(t)\right) dt$$

to be minimized.

The appropriate Hamiltonian is

$$H(x, z, u, p_1, p_2) = \tfrac{1}{2}(x^2 + u^2) + p_1 x + p_2(-x - z + u)$$

and the Euler–Lagrange equations for minimum cost are

$$\frac{dx}{dt} = x$$

$$\frac{dp_1}{dt} = -x - p_1 + p_2$$

$$\varepsilon\frac{dz}{dt} = -x - z + u$$

$$\varepsilon\frac{dp_2}{dt} = p_2$$

with the relation $\partial H/\partial u = 0$, or

$$u = -p_2,$$

and the terminal conditions

$$p_1(1) = x(1), \qquad p_2(1) = z(1).$$

The general solution of this constant coefficient system is

$$\begin{aligned}
x(t) &= \phantom{c_1(\varepsilon)e^{-t}} -2c_2(\varepsilon)(1+\varepsilon)e^t \\
p_1(t) &= c_1(\varepsilon)e^{-t} + c_2(\varepsilon)(1+\varepsilon)e^t + 2\varepsilon c_3(\varepsilon)e^{-(1-t)/\varepsilon} \\
z(t) &= 2c_2(\varepsilon)e^t - c_3(\varepsilon)(1+\varepsilon)e^{-(1-t)/\varepsilon} + c_4(\varepsilon)e^{-t/\varepsilon} \\
p_2(t) &= 2c_3(\varepsilon)(1+\varepsilon)e^{-(1-t)/\varepsilon},
\end{aligned}$$

where the $c_i(\varepsilon)$'s are arbitrary. Applying the boundary conditions, we have the optimal solution

$$x(t) = x_0 e^t$$

$$z(t) = \frac{x_0}{1+\varepsilon}\left[-e^t + e^{-t/\varepsilon} + \frac{e}{3}e^{-(1-t)/\varepsilon}\right] + z_0 e^{-t/\varepsilon} + O(e^{-1/\varepsilon})$$

$$u(t) = \frac{2}{3}\frac{x_0 e}{(1+\varepsilon)}e^{-(1-t)/\varepsilon} + O(e^{-1/\varepsilon})$$

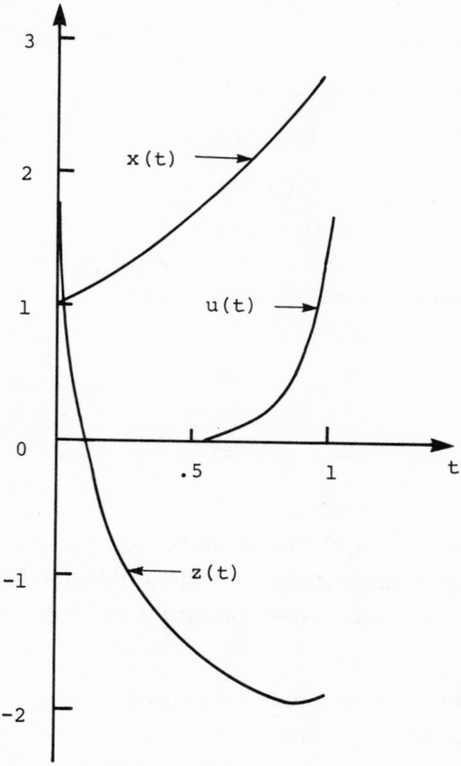

FIGURE 17 The solution of the control problem with $dx/dt = x$, $x(0) = 1$; $\varepsilon\, dz/dt = -x - z + u$, $z(0) = 2$; and $J(\varepsilon) = \frac{1}{2}(x^2(1) + \varepsilon z^2(1)) + \frac{1}{2}\int_0^1 (x^2(t) + u^2(t))\, dt$, $\varepsilon = 0.1$.

with the corresponding cost

$$J^*(\varepsilon) = \frac{x_0^2}{2}\left[e^2\left(1 + \frac{4\varepsilon}{9(1+\varepsilon)^2}\right) + \int_0^1 e^{2t}\,dt + \frac{4}{9}e^2\frac{\varepsilon}{(1+\varepsilon)^2}\int_0^\infty e^{-2\sigma}\,d\sigma\right]$$

$$= \frac{x_0^2}{4}\left(3e^2 - 1 + \varepsilon\frac{4e^2}{9(1+\varepsilon)^2}\right).$$

This solution is pictured in Fig. 17. It can also be simply obtained by the expansion procedure outlined above. Here, the reduced problem has the solution

$$X_0(t) = x_0 e^t$$

$$Z_0(t) = -x_0 e^t$$

$$U_0(t) = 0$$

with the corresponding cost

$$J_0^* = \frac{x_0^2}{4}(3e^2 - 1).$$

Remarks

1. In models for practical control problems, one must expect several small interrelated parameters to enter the problem in both regular and singular ways. Techniques appropriate for such problems could be developed as for the initial value problems in Chapter 4.

2. Linear state regulator problems for time invariant systems on infinite time intervals are important in applications. Controllability assumptions are needed [cf. Kalman *et al.* (1969)] and the singular perturbation problem is more subtle [cf. Hoppensteadt (1966)]. Considerable progress on this problem has been made using the Riccati

matrix formulation [cf. Kokotović and Yackel (1972) and O'Malley (1972b)]. Likewise, other regulator problems could be considered by proceeding analogously.

3. The method given here has been extended to certain nonlinear problems. Much progress has been made by Kokotović and his students [cf. Sannuti and Kokotović (1969), Kokotović and Sannuti (1968), Hadlock (1970), and Sannuti (1971)]. They also discuss models leading to such problems. O'Malley (1972d) further discusses some quasi-linear problems.

4. Note that the hypotheses made in solving the preceding problem involve systems of dimension less than $(2m + 2n) \times (2m + 2n)$. This order reduction is of great practical importance when working with large-order models. Alternative Riccati matrix approaches to the same problem involve additional computational advantages and feedback control [cf. Yackel and Wilde (1972) and Kung (1973)].

5. The limiting solution (X_0, Z_0, U_0, J_0^*) obtained is determined from the reduced problem (6.17) corresponding to the two-point problem for the system (6.9). It has not been shown that this limiting solution could be obtained from the reduced problem consisting of the system

$$\frac{dx}{dt} = A_1(t,0)x + A_2(t,0)z + B_1(t,0)u$$

$$0 = A_3(t,0)x + A_4(t,0)z + B_2(t,0)u,$$

the initial condition $x^0(0)$, and the cost functional $J(0)$. This alternative reduced problem seems to be the more natural one since it is immediately obtained. For certain problems, the equivalence of the two limiting problems has been shown [cf. Haddad and Kokotović (1971), O'Malley (1972b), Kokotović and Yackel (1972), and Kung (1973)]. The most general results, due to Sannuti, are unpublished. Problems without the ε multipliers of π_2 and π_3 in the terminal weighting matrix $\pi(\varepsilon)$ can also be solved [cf. Kung (1973)]. The natural reduced problem is not then appropriate, however.

6. The boundary layer behavior obtained in these control problems might be predicted by introduction of delta functions. We note that the function

$$\frac{a}{\varepsilon}e^{-at/\varepsilon}, \qquad 0 \le t \le 1, \quad a > 0$$

behaves like the delta function $\delta(t)$ as $\varepsilon \to 0$; i.e., for any differentiable $f(t)$,

$$\begin{aligned}\int_0^1 f(t)\frac{a}{\varepsilon}e^{-at/\varepsilon}\,dt &= f(0) - f(1)e^{-a/\varepsilon} + \int_0^1 f'(t)e^{-at/\varepsilon}\,dt \\ &= f(0) + O(\varepsilon).\end{aligned}$$

Likewise,

$$\frac{a}{\varepsilon}e^{-a(1-t)/\varepsilon}, \qquad 0 \le t \le 1, \quad a > 0$$

behaves like the delta function $\delta(1 - t)$ as $\varepsilon \to 0$. Recalling the preceding example, we note that the boundary layer behavior of the optimal control near $t = 1$ forced the nonuniform behavior of the resulting trajectory $z(t)$ there. This delta function formulation could be further utilized here and in other singular perturbation calculations.

7. Additional use of singular perturbations in optimal control problems includes the work of Lions (1972, 1973) on partial differential equations, Collins (1972) on bang-bang control, Jameson and O'Malley (1974) on singular arcs, and Bagirova *et al.* (1967).

CHAPTER 7

BOUNDARY VALUE PROBLEMS WITH MULTIPLE SOLUTIONS ARISING IN CHEMICAL REACTOR THEORY

Boundary value problems of the form

$$\begin{aligned} \varepsilon u'' + u' &= g(t,u), \qquad 0 \le t \le 1 \\ u'(0) - au(0) &= A \\ u'(1) + bu(1) &= B \end{aligned} \tag{7.1}$$

arise in the study of adiabatic tubular chemical flow reactors with axial diffusion [cf., e.g., Raymond and Amundson (1964), Burghardt and Zaleski (1968), and Cohen (1972)]. We shall be interested in obtaining asymptotic solutions of the problem as the positive parameter ε tends to zero. Physically, this means that the Peclet number is becoming large (as it would, for example, when the diffusivity becomes small). Mathematical aspects of the problem have been studied by Cohen and others. We shall take a, b, A, and B as constants and shall assume that $g(t, u)$ is infinitely differentiable in both arguments.

As $\varepsilon \to 0$, our experience with singular perturbation problems leads us to expect nonuniform convergence at $t = 0$ with convergence elsewhere to a solution of the reduced problem

$$U_0'(t) = g(t, U_0)$$
$$U_0'(1) + bU_0(1) = B,$$

if it exists. Thus $U_0(1)$ would satisfy the nonlinear equation

$$g(1, \alpha_0) + b\alpha_0 = B. \tag{7.2}$$

Corresponding to each root α_0 of (7.2), then, we shall assume that there is a unique solution of the corresponding nonlinear terminal value problem

$$\begin{aligned} U_0'(t) &= g(t, U_0) \\ U_0(1) &= \alpha_0 \end{aligned} \tag{7.3}$$

which exists throughout $0 \leq t \leq 1$.

We shall then seek asymptotic solutions $u(t, \varepsilon)$ of (7.1) converging to the solution $U_0(t)$ of (7.3) for $t > 0$. We shall give a detailed discussion of the three cases: (a) α_0 is a simple root of (7.2), (b) α_0 is a double root of (7.2), and (c) α_0 is a triple root of (7.2). The general case where α_0 is a root of arbitrary finite multiplicity follows analogously [cf. Chen (1972)].

CASE a: In Case a, α_0 is a simple root of (7.2) so

$$P(\alpha_0) \equiv g(1, \alpha_0) + b\alpha_0 - B = 0 \tag{7.4a}$$

while

$$P'(\alpha_0) = g_u(1, \alpha_0) + b \neq 0 \tag{7.4b}$$

and we shall seek a solution $u(t, \varepsilon)$ of (7.1) of the form

$$u(t, \varepsilon) = U(t, \varepsilon) + \varepsilon v(\tau, \varepsilon), \tag{7.5}$$

where the outer solution is expanded as

$$U(t, \varepsilon) \sim \sum_{j=0}^{\infty} U_j(t)\varepsilon^j$$

and the boundary layer correction as

$$v(\tau, \varepsilon) \sim \sum_{j=0}^{\infty} v_j(\tau)\varepsilon^j$$

with the coefficients $v_j(\tau)$ tending to zero as the boundary layer coordinate

$$\tau = t/\varepsilon \tag{7.6}$$

tends to infinity. If this scheme is possible, $u(t, \varepsilon)$ will converge to $U_0(t)$ as $\varepsilon \to 0$ throughout $0 \leq t \leq 1$ while $u'(t, \varepsilon)$ will converge nonuniformly at $t = 0$. Thus, the outer solution $U(t, \varepsilon)$ must satisfy the differential equation and the terminal boundary condition of (7.1), i.e.,

$$\varepsilon U'' + U' = g(t, U)$$
$$U'(1, \varepsilon) + bU(1, \varepsilon) = B.$$

These equations are automatically satisfied when $\varepsilon = 0$ since $U_0(t)$ satisfies the reduced problem. They imply that higher-order coefficients $U_j(t)$ with $j > 0$ must satisfy linear terminal value problems

$$\begin{aligned} U_j'(t) &= g_u(t, U_0)U_j + \tilde{g}_{j-1}(t) \\ U_j'(1) + bU_j(1) &= 0 \end{aligned} \tag{7.7}$$

on $0 \leq t \leq 1$, where $\tilde{g}_{j-1}$ is known successively. Thus, the terminal value $U_j(1)$ is uniquely determined by the linear equation

$$P'(\alpha_0)\, U_j(1) = \Big(g_u(1, \alpha_0) + b\Big)\, U_j(1) = -\tilde{g}_{j-1}(1) \tag{7.8}$$

when α_0 is a simple root of (7.2). When $P'(\alpha_0) = 0$, we generally encounter a difficulty, and the outer expansion cannot be obtained as a power series in ε. In Case a, however, the outer expansion $U(t, \varepsilon)$ is uniquely and straightforwardly determined.

The boundary layer correction $v(\tau, \varepsilon)$ must account for the nonuniform convergence near $t = 0$. Since the outer solution is known, the boundary layer correction must satisfy the initial value problem

$$\begin{aligned} v_{\tau\tau} + v_\tau &= g\Big(\varepsilon\tau, U(\varepsilon\tau, \varepsilon) + \varepsilon v(\tau, \varepsilon)\Big) - g\Big(\varepsilon\tau, U(\varepsilon\tau, \varepsilon)\Big) \\ v_\tau(0, \varepsilon) &= A - U'(0, \varepsilon) + aU(0, \varepsilon) + \varepsilon a v(0, \varepsilon) \end{aligned} \tag{7.9}$$

and v must tend to zero as τ tends to infinity. When $\varepsilon = 0$, then,

$$\begin{aligned} v_{0\tau\tau} + v_{0\tau} &= 0 \\ v_{0\tau}(0) &= A - U_0'(0) + aU_0(0). \end{aligned}$$

Thus, we have

$$v_0(\tau) = -v_{0\tau}(0)e^{-\tau}. \tag{7.10}$$

Similarly, higher-order coefficients $v_j(\tau)$ must satisfy

$$\begin{aligned} v_{j\tau\tau} + v_{j\tau} &= \tilde{G}_{j-1}(\tau) \\ v_{j\tau}(0) &= \tilde{\tilde{B}}_{j-1}, \end{aligned}$$

where $\tilde{G}_{j-1}$ and $\tilde{\tilde{B}}_{j-1}$ are known successively and $\tilde{G}_{j-1}(\tau) = O(e^{-(1-\delta)\tau})$ for any $\delta > 0$ as $\tau \to \infty$. Thus,

$$\begin{aligned} v_j(\tau) &= -\tilde{\tilde{B}}_{j-1}\, e^{-\tau} - \int_\tau^\infty \int_0^r e^{-(r-s)}\, \tilde{G}_{j-1}(s)\, ds\, dr \\ &= O(e^{-(1-\delta)\tau}) \quad \text{as} \quad \tau \to \infty; \end{aligned} \tag{7.11}$$

i.e., the boundary layer correction $v(\tau, \varepsilon)$ can be uniquely determined termwise with coefficients tending to zero as τ tends to infinity.

CASE b: Now α_0 is a double root of (7.2) so

$$P(\alpha_0) = g(1, \alpha_0) + b\alpha_0 - B = 0 \tag{7.12a}$$

$$P'(\alpha_0) = g_u(1, \alpha_0) + b = 0 \tag{7.12b}$$

and

$$P''(\alpha_0) = g_{uu}(1, \alpha_0) \neq 0 \tag{7.12c}$$

and we know that an outer expansion $U(t, \varepsilon)$ cannot generally be determined as a power series in ε. We avoid the possible problem by seeking an outer expansion of the form

$$U(t, \sqrt{\varepsilon}) \sim \sum_{j=0}^{\infty} U_j(t)(\sqrt{\varepsilon})^j, \tag{7.13}$$

where $\sqrt{\varepsilon} > 0$ and $U_0(t)$ again satisfies (7.3) throughout $0 \leq t \leq 1$. Since the outer solution must satisfy the differential equation and the terminal boundary condition of (7.1), successive coefficients of $(\sqrt{\varepsilon})^j$ imply that

$$\begin{aligned}
&U_1'(t) = g_u(t, U_0)U_1 \\
&U_1'(1) + bU_1(1) = 0, \\
&U_2'(t) = g_u(t, U_0)U_2 + \tfrac{1}{2}g_{uu}(t, U_0)U_1^2 - U_0'' \\
&U_2'(1) + bU_2(1) = 0, \\
&U_3'(t) = g_u(t, U_0)U_3 + g_{uu}(t, U_0)U_1 U_2 \\
&\qquad\qquad + \frac{1}{3!}g_{uuu}(t, U_0)U_1^3 - U_1'' \\
&U_3'(1) + bU_3(1) = 0,
\end{aligned}$$

etc. By (7.12), note that $(g_u(1, \alpha_0) + b)U_1(1) = 0$, so the terminal condition for U_1 is satisfied whatever $U_1(1)$. Likewise the boundary condition for U_2 becomes

$$g_{uu}(1, \alpha_0)\, U_1^2(1) = 2U_0''(1). \tag{7.14}$$

Thus, to obtain a real value $U_1(1)$ we ask that

$$U_0''(1)\, g_{uu}(1, \alpha_0) > 0 \tag{7.15}$$

[anticipating a later difficulty if $U_0''(1) = 0$]. Under this assumption, two distinct terminal values $U_1(1)$ are possible: namely $U_1(1) = \pm\alpha_1$, where $\alpha_1 = [2U_0''(1)/g_{uu}(1, \alpha_0)]^{1/2} > 0$. Thus, there are two determinations of $U_1(t)$ as unique solutions of the linear problems

$$\begin{aligned} U_1'(t) &= g_u(t, U_0)U_1 \\ U_1(1) &= \pm\alpha_1; \end{aligned} \tag{7.16}$$

i.e., $U_1(t) = \pm\alpha_1 \exp[\int_1^t g_u(s, U_0(s))\, ds]$. Continuing, the boundary condition for U_3 becomes

$$g_{uu}(1, \alpha_0)\, U_1(1)\, U_2(1) + \frac{1}{3!} g_{uuu}(1, \alpha_0)\, U_1^3(1) - U_1''(1) = 0.$$

Since (7.15) implies that $g_{uu}(1, \alpha_0)\, U_1(1) \neq 0$, this equation yields a unique determination of $U_2(1) = \alpha_2$. Thus, U_2 satisfies the linear problem

$$\begin{aligned} U_2'(t) &= g_u(t, U_0)U_2 + \tfrac{1}{2} g_{uu}(t, U_0)\, U_1^2 - U_0'' \\ U_2(1) &= \alpha_2 \equiv g(1, \alpha_0) + \frac{1}{g_{uu}(1, \alpha_0)} \left[g_u^2(1, \alpha_0) - \frac{1}{6} g_{uuu}(1, \alpha_0)\, \alpha_1^2 \right]. \end{aligned} \tag{7.17}$$

Continuing, we can successively determine all initial values $U_j(1)$ and higher-order coefficients $U_j(t)$ uniquely once $U_1(1)$ is selected. Thus, two outer solutions $U(t, \sqrt{\varepsilon})$ can be formally obtained corresponding to the double root α_0 of (7.2) which satisfies the restriction (7.15).

Corresponding to each of these outer solutions $U(t, \sqrt{\varepsilon})$, we shall seek an asymptotic solution of the form

$$u(t, \varepsilon) = U(t, \sqrt{\varepsilon}) + \varepsilon v(\tau, \sqrt{\varepsilon}), \tag{7.18}$$

where the boundary layer correction $v(\tau, \sqrt{\varepsilon})$ has the asymptotic expansion

$$v(\tau, \sqrt{\varepsilon}) \sim \sum_{j=0}^{\infty} v_j(\tau)(\sqrt{\varepsilon})^j$$

and the coefficients $v_j \to 0$ as $\tau = t/\varepsilon \to \infty$. It follows that $v(\tau, \sqrt{\varepsilon})$ must satisfy the initial value problem

$$\begin{aligned} v_{\tau\tau} + v_\tau &= g\left(\varepsilon\tau, U(\varepsilon\tau, \sqrt{\varepsilon}) + \varepsilon v(\tau, \sqrt{\varepsilon})\right) - g\left(\varepsilon\tau, U(\varepsilon\tau, \sqrt{\varepsilon})\right) \\ v_\tau(0, \sqrt{\varepsilon}) &= A - U'(0, \sqrt{\varepsilon}) + aU(0, \sqrt{\varepsilon}) + \varepsilon a v(0, \sqrt{\varepsilon}). \end{aligned} \tag{7.19}$$

The coefficients in the expansion for v can be uniquely obtained as in Case a.

CASE c: Here α_0 is a triple root of (7.2), so

$$P(\alpha_0) = g(1, \alpha_0) + b\alpha_0 - B = 0 \tag{7.20a}$$

$$P'(\alpha_0) = g_u(1, \alpha_0) + b = 0 \tag{7.20b}$$

$$P''(\alpha_0) = g_{uu}(1, \alpha_0) = 0 \tag{7.20c}$$

$$P'''(\alpha_0) = g_{uuu}(1, \alpha_0) \neq 0, \tag{7.20d}$$

and we will attempt to obtain an outer expansion of the form

$$U(t, \varepsilon^{1/3}) \sim \sum_{j=0}^{\infty} U_j(t)(\varepsilon^{j/3}), \tag{7.21}$$

where U_0 satisfies (7.3) and $\varepsilon^{1/3} > 0$. The differential equation and the terminal condition of (7.1) then imply that

$$U_1'(t) = g_u(t, U_0)U_1$$

$$U_1'(1) + bU_1(1) = 0,$$

$$U_2'(t) = g_u(t, U_0)U_2 + \tfrac{1}{2}g_{uu}(t, U_0)U_1^2$$

$$U_2'(1) + bU_2(1) = 0,$$

$$\begin{aligned} U_3'(t) = {} & g_u(t, U_0)U_3 + g_{uu}(t, U_0)U_1 U_2 \\ & + \frac{1}{3!}g_{uuu}(t, U_0)\, U_1^3 - U_0'' \end{aligned}$$

$$U_3'(1) + bU_3(1) = 0,$$

$$\begin{aligned} U_4'(t) = {} & g_u(t, U_0)U_4 + g_{uu}(t, U_0)(U_1 U_3 + \tfrac{1}{2}U_2^2) \\ & + \tfrac{1}{2}g_{uuu}(t, U_0)\, U_1^3 U_2 + \frac{1}{4!}g_{uuuu}(t, U_0)\, U_1^4 - U_1'' \end{aligned}$$

$$U_4'(1) + bU_4(1) = 0,$$

etc. Note that (7.20) implies that the terminal conditions for U_1 and U_2 are satisfied whatever $U_1(1)$ and $U_2(1)$. Also the boundary condition for U_3 becomes

$$\frac{1}{3!} g_{uuu}(1, \alpha_0) U_1^3(1) - U_0''(1) = 0. \tag{7.22}$$

Thus, we ask that

$$U_0''(1) \neq 0 \tag{7.23}$$

and we obtain only one real nonzero determination $U_1(1) = \alpha_1$. Then U_1 is the unique solution of

$$\begin{aligned} U_1'(t) &= g_u(t, U_0) U_1 \\ U_1(1) &= \alpha_1 = \left(\frac{6U_0''(1)}{g_{uuu}(1, \alpha_0)} \right)^{1/3}. \end{aligned} \tag{7.24}$$

Continuing, the terminal condition for U_4 becomes

$$\frac{1}{2} g_{uuu}(1, \alpha_0) U_1^3(1) U_2(1) + \frac{1}{4!} g_{uuuu}(1, \alpha_0) U_1^4(1) - U_1''(1) = 0. \tag{7.25}$$

Since (7.25) implies that $U_1(1) \neq 0$, we can uniquely determine $U_2(1) = \alpha_2$ and U_2 must satisfy

$$\begin{aligned} U_2'(t) &= g_u(t, U_0) U_2 + \tfrac{1}{2} g_{uu}(t, U_0) U_1^2 \\ U_2(1) &= \alpha_2. \end{aligned} \tag{7.26}$$

Continuing, we can uniquely obtain higher-order coefficients $U_j(t)$ in a straightforward fashion.

Here we obtain the behavior near $t = 0$ by seeking an asymptotic solution $u(t, \varepsilon)$ such that

$$u(t, \varepsilon) = U(t, \varepsilon^{1/3}) + \varepsilon v(\tau, \varepsilon^{1/3}), \tag{7.27}$$

where the boundary layer correction $v(\tau, \varepsilon^{1/3})$ has an asymptotic expansion

$$v(\tau, \varepsilon^{1/3}) \sim \sum_{j=0}^{\infty} v_j(\tau) \varepsilon^{j/3}$$

such that the $v_j(\tau) \to 0$ as $\tau = t/\varepsilon \to \infty$.

Though α_0 is a root of (7.2) of multiplicity three, satisfying (7.23), note that we have only obtained a single corresponding asymptotic solution of (7.1).

Generalizing slightly, we have

THEOREM 7: *Consider the boundary value problem*

$$\varepsilon u'' + u' = g(t, u), \qquad 0 \leq t \leq 1$$

$$u'(0) - au(0) = A$$

$$u'(1) + bu(1) = B.$$

Let α_0 satisfy the nonlinear equation

$$P(\alpha_0) = g(1, \alpha_0) + b\alpha_0 - B = 0$$

and suppose that the reduced problem

$$U_0'(t) = g(t, U_0)$$

$$U_0(1) = \alpha_0$$

has a unique solution throughout $0 \leq t \leq 1$. Then, for ε sufficiently small,

(a) *corresponding to every (simple) root α_0 of multiplicity $m = 1$, the boundary value problem has a solution $u(t, \varepsilon)$.*

(b) *corresponding to every α_0 of even multiplicity m such that*

$$U_0''(1) \frac{\partial^m g(1, \alpha_0)}{\partial u^m} > 0, \tag{7.28}$$

there are two distinct solutions $u(t, \varepsilon)$.

(c) *corresponding to every root α_0 of odd multiplicity $m > 1$ such that*

$$U_0''(1) \neq 0, \tag{7.29}$$

there is one solution $u(t, \varepsilon)$.

In all cases, the solution $u(t, \varepsilon)$ *is of the form*

$$u(t, \varepsilon) = U(t, \varepsilon^{1/m}) + \varepsilon v(t/\varepsilon, \varepsilon^{1/m});$$

i.e., for every integer $N \geq m$,

$$u(t, \varepsilon) = \sum_{j=0}^{m-1} U_j(t)\varepsilon^{j/m} + \sum_{j=m}^{N} \left(U_j(t) + v_{j-m}(t/\varepsilon)\right)\varepsilon^{j/m} + \varepsilon^{(N+1)/m} R(t, \varepsilon). \tag{7.30}$$

where $R(t, \varepsilon)$ *is bounded throughout* $0 < t < 1$ *and the* v_{j-m} *'s* $\rightarrow 0$ *as* $t/\varepsilon \rightarrow \infty$.

Remarks

1. The extra hypotheses (7.28) and (7.29) on $U_0''(1)$ when $m > 1$ allow us to solve the boundary condition

$$U_m'(1) + bU_m(1) = \frac{1}{m!} U_1^m(1) \frac{\partial^m g(1, \alpha_0)}{\partial u^m} - U_0''(1) = 0$$

for $U_1(1)$ and later boundary conditions for successive $U_j(1)$'s since

$$\frac{\partial^m g(1, \alpha_0)}{\partial u^m} \neq 0.$$

The results of Parter (1972) indicate that such conditions may be necessary for existence.

2. The proof of asymptotic correctness follows simply from the previous results for nonlinear initial value problems (cf. Chapter 4). It has not been shown, however, that the asymptotic solutions constructed are the only solutions existing. Finally note that all these solutions converge to a solution $U_0(t)$ of the reduced problem throughout $0 \leq t \leq 1$ while derivatives generally converge nonuniformly as $\varepsilon \rightarrow 0$ at $t = 0$.

3. Analogous results have been obtained for the more general problem

$$\varepsilon u'' + f(u, t, \varepsilon)u' = g(u, t, \varepsilon), \qquad 0 \le t \le 1$$
$$m\big(u(0), u'(0), \varepsilon\big) = 0$$
$$n\big(u(1), u'(1), \varepsilon\big) = 0$$

[see O'Malley (1972c)]. For these problems, multiple solutions for the boundary layer correction (corresponding to any outer solution) are also possible. Generalizations to two-parameter problems and vector equations are discussed by Chen (1972). Keller (1973) discusses the stability problem for the diffusion equation for which the ordinary differential equation (7.1) provides the steady state limit. A number of related problems in chemical reactor theory are discussed by Cohen (1972a). We note that the techniques developed here for problem (7.1) and the resulting multiple solutions may actually be more relevant to batch reactors and to tubular reactors with recycle than to (7.1) itself.

CHAPTER 8

SOME TURNING POINT PROBLEMS

1. A SIMPLE PROBLEM

Consider the special linear boundary value problem

$$\varepsilon y'' + 2\alpha x y' - \alpha\beta y = 0, \qquad -1 \leq x \leq 1$$
$$\text{with } y(\pm 1) \text{ being prescribed} \tag{8.1}$$

for α and β constants with $\alpha \neq 0$ and for ε a small positive parameter. Note that the results of Section 3.1 are not applicable here since

$$a(x) = 2\alpha x$$

has a zero within $[-1, 1]$. Further, note that the reduced problem obtained by setting $\varepsilon = 0$ in (8.1) is singular at $x = 0$. Near x

$= -1(+1)$, however, $a(x)$ is negative (positive) if $\alpha > 0$, so those earlier results lead us to expect uniform convergence to occur at both endpoints when $\alpha > 0$. On the other hand, if $\alpha < 0$, $a(x)$ is positive (negative) near the left (right) endpoint, so we then expect nonuniform convergence at both $x = \pm 1$. Thus, one might expect the limiting solution for $\alpha > 0$ to satisfy

$$2xz_1' - \beta z_1 = 0, \qquad z_1(-1) = y(-1) \qquad \text{on} \quad [-1, 0)$$

and

$$2xz_2' - \beta z_2 = 0, \qquad z_2(1) = y(1) \qquad \text{on} \quad (0, 1],$$

with possibly unbounded behavior at the "turning point" $x = 0$, depending on the sign of β. Likewise, one might argue that the limiting solution within $(-1, 1)$ for $\alpha < 0$ should be the trivial solution of the reduced equation. These conjectures are correct except for a countable number of β values when a "resonance" phenomenon occurs.

The problem (8.1) can, fortunately, be solved in terms of the appropriate Weber (or parabolic cylinder) functions [cf. Whittaker and Watson (1952)]. [In certain very special cases (e.g., when $\beta = 0$) the equation may be simply integrated and special functions need not be used.] Note that, if y satisfies (8.1), $u = y \exp(\alpha x^2/2\varepsilon)$ satisfies

$$\varepsilon^2 \frac{d^2 u}{dx^2} = \big(\alpha^2 x^2 + \varepsilon\alpha(1 + \beta)\big)u$$

and, since the parabolic cylinder functions $D_n(t)$ and $D_{n-1}(it)$ are linearly independent solutions of

$$\frac{d^2 w}{dt^2} + (n + \tfrac{1}{2} - \tfrac{1}{4}t^2)w = 0,$$

the general solution of the differential equation of (8.1) is

$$y(x) = \exp\left(-\frac{\alpha x^2}{2\varepsilon}\right)\left[C_1 D_{-1-\frac{1}{2}\beta}\left(\left(\frac{2\alpha}{\varepsilon}\right)^{1/2} x\right) + C_2 D_{\frac{1}{2}\beta}\left(i\left(\frac{2\alpha}{\varepsilon}\right)^{1/2} x\right)\right] \tag{8.2}$$

for arbitrary C_1 and C_2. For (8.1), then,

$$y(\pm 1) = \exp\left(-\frac{\alpha}{2\varepsilon}\right)\left[C_1 D_{-1-\frac{1}{2}\beta}\left(\pm\left(\frac{2\alpha}{\varepsilon}\right)^{1/2}\right)\right.$$
$$\left. + C_2 D_{\frac{1}{2}\beta}\left(\pm i\left(\frac{2\alpha}{\varepsilon}\right)^{1/2}\right)\right] \tag{8.3}$$

provide two linear equations to determine C_1 and C_2. To solve these equations for small ε, we need the following asymptotic approximations for $D_n(z)$:

$$D_n(z) = \begin{cases} \exp\left(-\frac{z^2}{4}\right) z^n \left(1 + o(1)\right), & \text{as } |z| \to \infty, \quad |\arg z| < \frac{3\pi}{4} \\ \exp\left(-\frac{z^2}{4}\right) z^n \left(1 + o(1)\right) \\ \quad - \frac{(2\pi)^{1/2}}{\Gamma(-n)} \frac{e^{n\pi i} \exp(z^2/4)}{z^{n+1}} \left(1 + o(1)\right) \\ \qquad \text{as } |z| \to \infty, \quad \frac{\pi}{4} < \arg z < \frac{5\pi}{4}. \end{cases} \tag{8.4}$$

Note that $D_n(z)$ is exponentially large when $|z| \to \infty$ and $\arg z = \pi$ unless n is a nonnegative integer. Then

$$D_n(z) = \exp\left(-\frac{z^2}{4}\right) He_n(z), \qquad n = 0, 1, 2, \ldots,$$

where He_n, the nth Hermite polynomial, satisfies

$$He_n(z) = z^n\left(1 + o(1)\right) \qquad \text{as} \quad |z| \to \infty.$$

Thus, when n is a nonnegative integer, $D_n(z)$ is exponentially small as $|z| \to \infty$ with $\arg z = \pi$. This difference has a substantial influence on the behavior of the asymptotic solutions to (8.1).

For this reason, we shall separately consider four cases.

CASE (*i*): $\alpha > 0$, $\beta \neq -2n$, $n = 1, 2, \ldots$. Using the appropriate asymptotic expansions from (8.4), the linear equations (8.3) have a unique solution with the asymptotic determinations

$$C_1(\varepsilon) = \frac{e^{-\alpha/2\varepsilon}}{D_{-1-\frac{1}{2}\beta}\left(-\left(\frac{2\alpha}{\varepsilon}\right)^{1/2}\right)}[y(-1) - (-1)^{\beta/2}y(1) + o(1)]$$

and

$$C_2(\varepsilon) = \frac{e^{-\alpha/2\varepsilon}}{D_{\frac{1}{2}\beta}\left(-i\left(\frac{2\alpha}{\varepsilon}\right)^{1/2}\right)}[(-1)^{\beta/2}y(1) + o(1)].$$

Thus, (8.2)–(8.4) imply that for $x > 0$,

$$y(x) \sim \frac{\Gamma(-1 - \beta/2)}{2\left(\frac{\pi\alpha}{\varepsilon}\right)^{1/2}}\Big(y(-1) - (-1)^{\beta/2}y(1) + o(1)\Big)\frac{\exp(-\alpha x^2/\varepsilon)}{x^{1+\beta/2}}$$
$$+ \Big((-1)^{\beta/2}y(1) + o(1)\Big)(-x)^{\beta/2}$$

or, for any $\delta > 0$,

$$y(x) = O\Big(\exp[-\alpha x^2(1 - \delta)/\varepsilon]\Big) + \Big(y(1)x^{\beta/2} + o(1)\Big),$$

so

$$y(x) \to z_2(x) \equiv y(1)x^{\beta/2} \quad \text{as} \quad \varepsilon \to 0 \quad \text{for} \quad x > 0. \tag{8.5}$$

Likewise, for $x < 0$,

$$y(x) = (-x)^{\beta/2}\Big[\Big(y(-1) - (-1)^{\beta/2}y(1) + o(1)\Big)$$
$$+ \Big((-1)^{\beta/2}y(1) + o(1)\Big)\Big]$$

or

$$y(x) \to z_1(x) \equiv y(-1)(-x)^{\beta/2} \quad \text{as} \quad \varepsilon \to 0 \quad \text{for} \quad x < 0. \tag{8.6}$$

Finally

$$y(0) = \left(-\left(\frac{\varepsilon}{2\alpha}\right)^{1/2}\right)^{\beta/2}\left[\frac{(y(-1) - (-1)^{\beta/2}y(1))}{2^{1+\beta/4}}\frac{\Gamma(1 + \beta/2)}{\Gamma(1 + \beta/4)}\right.$$

$$\left. + \frac{\sqrt{\pi}\,y(1)(-i2)^{\beta/4}}{\Gamma(\frac{1}{2} - \beta/4)} + o(1)\right] = O(\varepsilon^{\beta/4}) \tag{8.7}$$

since

$$D_n(0) = \frac{\sqrt{\pi}\,2^{n/2}}{\Gamma((1 - n)/2)}.$$

Thus, if $\beta > 0$, y converges uniformly on $[-1, 1]$ since $y(0) \to 0$ and $z_1(0) = z_2(0) = 0$. (Derivatives of y may still converge nonuniformly at $x = 0$, however.) If $\beta = 0$ and $y(1) \neq y(-1)$, convergence is nonuniform at $x = 0$ since

$$y(0) \to \tfrac{1}{2}\big(y(1) + y(-1)\big),$$

while $z_1(x) \equiv y(-1)$ and $z_2(x) \equiv y(1)$. Lastly, if $\beta < 0$, but not an even integer, $z_1(0)$ and $z_2(0)$ are both undefined while $y(0)$ becomes unbounded like $\varepsilon^{\beta/4}$. In summary, then, the anticipated convergence to the solution of the appropriate reduced problem occurs on the intervals $-1 \leq x < 0$ and $0 < x \leq 1$.

Note that complete asymptotic expansions for the solution y of (8.1) would be obtained if we used complete expansions for the Weber functions and the coefficients C_i in (8.2).

CASE (*ii*): $\alpha < 0$, $\beta \neq 2m$, $m = 0, 1, 2, \ldots$. Here (8.3) and (8.4) imply that

$$C_1 = \frac{e^{\alpha/2\varepsilon}}{D_{-1-\beta/2}\left(i\left(\frac{-2\alpha}{\varepsilon}\right)^{1/2}\right)}\big(-(-1)^{\beta/2}y(-1) + o(1)\big)$$

and

$$C_2 = \frac{e^{\alpha/2\varepsilon}}{D_{\beta/2}\left(-\left(\frac{-2\alpha}{\varepsilon}\right)^{1/2}\right)}\big(y(1) + (-1)^{\beta/2}y(-1) + o(1)\big).$$

Thus, (8.2)–(8.4) imply that

$$y(x) = \frac{1}{x^{1+\beta/2}}\big(y(1) + o(1)\big)\exp[\alpha(1 - x^2)/\varepsilon] \qquad \text{for} \quad x > 0,$$

$$y(x) = \frac{1}{(-x)^{1+\beta/2}}\big(y(-1) + o(1)\big)\exp[\alpha(1 - x^2)/\varepsilon] + O(e^{\alpha(1-\delta)/\varepsilon})$$

$$\text{for} \quad x < 0,$$

and

$$y(0) = O(e^{\alpha(1-\delta)/\varepsilon})$$

for any $\delta > 0$. Since $\alpha < 0$, it follows that

$$y \to 0 \qquad \text{within} \quad (-1, 1) \quad \text{as} \quad \varepsilon \to 0 \tag{8.8}$$

with nonuniform convergence generally occurring at both endpoints.

CASE (iii): $\alpha < 0$, $\beta = 2m$, $m = 0, 1, 2, \ldots$. Here (8.2) becomes

$$y(x) = C_1 \exp\left(-\frac{\alpha x^2}{2\varepsilon}\right) D_{-1-m}\left(i\left(-\frac{2\alpha}{\varepsilon}\right)^{1/2} x\right) + C_2 He_m\left(-\left(-\frac{2\alpha}{\varepsilon}\right)^{1/2} x\right)$$

and

$$C_1 = \frac{e^{-\alpha/2\varepsilon}}{D_{-1-m}\left(i\left(-\frac{2\alpha}{\varepsilon}\right)^{1/2}\right)}\left[\tfrac{1}{2}\big(y(1) - (-1)^m y(-1)\big) + o(1)\right]$$

while

$$C_2 = \frac{1}{He_m\left(-\left(-\frac{2\alpha}{\varepsilon}\right)^{1/2}\right)}\left[\tfrac{1}{2}\big(y(1) + (-1)^m y(-1)\big) + o(1)\right].$$

Thus, for $x \neq 0$,

$$y(x) = \tfrac{1}{2}\Big[\Big(y(1) - (-1)^m y(-1)\Big)\frac{\exp[\alpha(1-x^2)/\varepsilon]}{x^{1+m}} + \Big(y(1) + (-1)^m y(-1)\Big)x^m\Big] + o(1)$$

and

$$y(0) = O(e^{\alpha(1-\delta)/\varepsilon}) \qquad \text{for any} \quad \delta > 0.$$

Since $\alpha < 0$,

$$y(x) \to \tfrac{1}{2}\Big(y(1) + (-1)^m y(-1)\Big)x^m \qquad \text{as} \quad \varepsilon \to 0 \tag{8.9}$$

away from $x = \pm 1$ where nonuniform convergence generally occurs. Note that, again, the limiting solution satisfies the reduced equation. Here, however, the limiting solution is nontrivial unless

$$y(1) = (-1)^{m-1} y(-1)$$

CASE (iv): $\alpha > 0$, $\beta = -2n$, $n = 1, 2, \ldots$. Here (8.2) has the form

$$y(x) = C_1 \exp\Big(-\frac{\alpha x^2}{\varepsilon}\Big) He_{-1+n}\Big(\Big(\frac{2\alpha}{\varepsilon}\Big)^{1/2} x\Big) + C_2 \exp\Big(-\frac{\alpha x^2}{2\varepsilon}\Big) D_{-n}\Big(i\Big(\frac{2\alpha}{\varepsilon}\Big)^{1/2} x\Big),$$

where

$$C_1 = \frac{e^{\alpha/\varepsilon}}{He_{-1+n}((2\alpha/\varepsilon)^{1/2})}\Big[\tfrac{1}{2}\Big(y(1) - (-1)^n y(-1)\Big) + o(1)\Big]$$

and

$$C_2 = \frac{e^{\alpha/2\varepsilon}}{D_{-n}(i(2\alpha/\varepsilon)^{1/2})}\Big[\tfrac{1}{2}\Big(y(1) + (-1)^n y(-1)\Big) + o(1)\Big].$$

Thus, for $x \neq 0$,

$$y(x) = \frac{\exp[\alpha(1 - x^2)/\varepsilon]}{x^{1-n}}\left[\tfrac{1}{2}\Big(y(1) - (-1)^n y(-1)\Big) + o(1)\right]$$

$$+ x^{-n}\left[\tfrac{1}{2}\Big(y(1) + (-1)^n y(-1)\Big) + o(1)\right]$$

or

$$y(x) = O(\exp[\alpha(1 - x^2)/\varepsilon]) \tag{8.10}$$

so y becomes exponentially large as $\varepsilon \to 0$ away from $x = \pm 1$. This unpleasant behavior can be illustrated by the special equation

$$\varepsilon y'' + 2xy' + 2y = 0,$$

which can be directly integrated without use of special functions.

We summarize our results in Table 1.

TABLE 1

Case	Limiting solution as $\varepsilon \to 0$
(i) $\alpha > 0$, $\beta \neq -2, -4, -6, \ldots$	$y(-1)(-x)^{\beta/2}$, $-1 \leq x < 0$ $O(\varepsilon^{\beta/4})$, $x = 0$ $y(1)x^{\beta/2}$, $0 < x \leq 1$
(ii) $\alpha < 0$, $\beta \neq 0, 2, 4, \ldots$	0, $-1 < x < 1$
(iii) $\alpha < 0$, $\beta = 0, 2, 4, \ldots$	$\tfrac{1}{2}(y(1)x^{\beta/2} + y(-1)(-x)^{\beta/2})$, $-1 < x < 1$
(iv) $\alpha > 0$, $\beta = -2, -4, -6, \ldots$	Solution becomes exponentially large within $(-1, 1)$.

The limiting solution of

$$\varepsilon y'' + 2\alpha xy' - \alpha\beta y = 0, \qquad -1 \leq x \leq 1$$

$$y(\pm 1) \quad \text{prescribed}$$

is given in the table.

Examples illustrating Cases (i)–(iv) are given in Figs. 18–22.

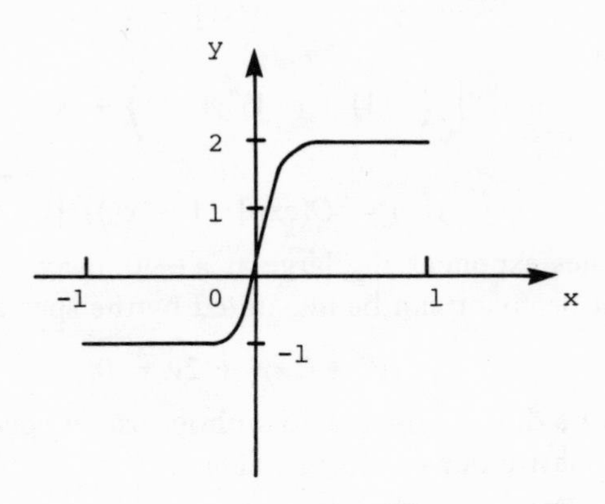

FIGURE 18 The solution of $\varepsilon y'' + 2xy' = 0$, $y(-1) = -1$, $y(1) = 2$, ε small.

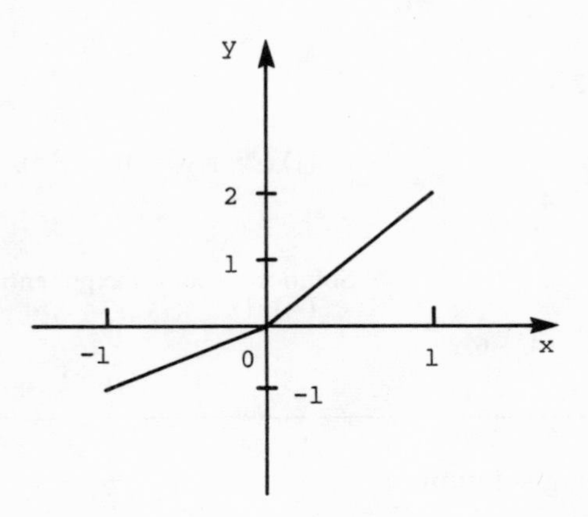

FIGURE 19 The solution of $\varepsilon y'' + 2xy' - 2y = 0$, $y(-1) = -1$, $y(1) = 2$, ε small.

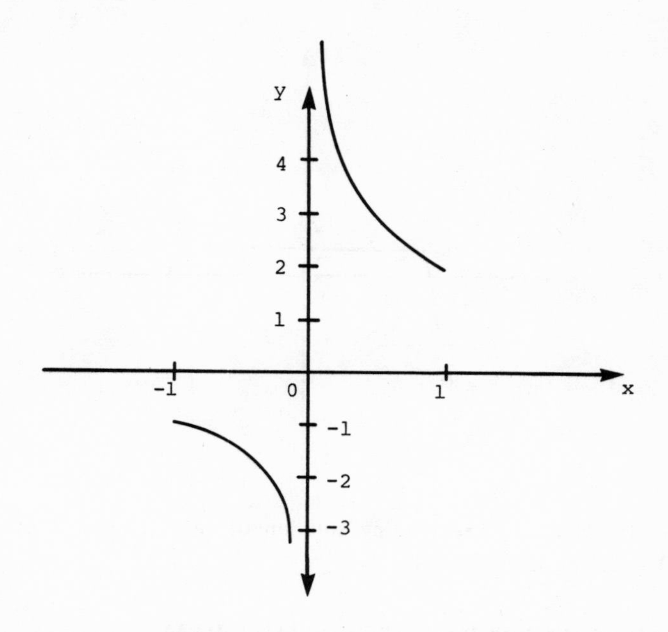

FIGURE 20 The solution of $\varepsilon y'' + 2xy' + y = 0$, $y(-1) = -1$, $y(1) = 2$, ε small.

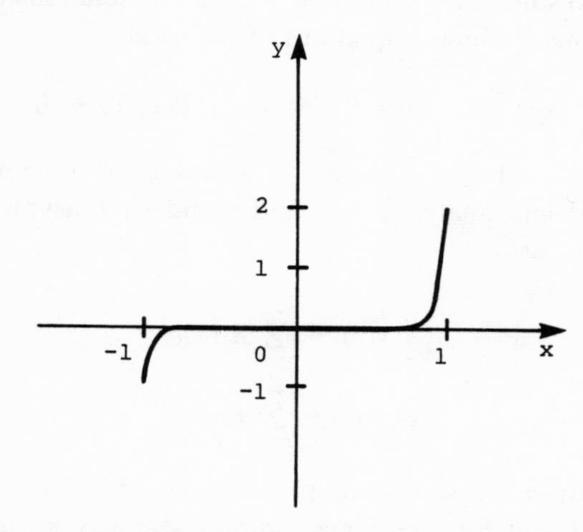

FIGURE 21 The solution of $\varepsilon y'' - 2xy' + y = 0$, $y(-1) = -1$, $y(1) = 2$, ε small.

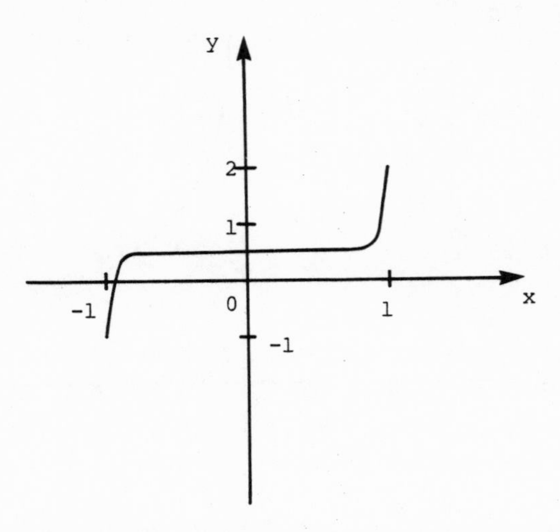

FIGURE 22 The limiting solution of $\varepsilon y'' - 2xy' = 0$, $y(-1) = -1$, $y(1) = 2$, ε small.

2. A UNIFORM REDUCTION THEOREM

We wish to study the asymptotic behavior of solutions to boundary value problems for linear equations of the form

$$\varepsilon y'' + 2x\ A(x, \varepsilon)y' - A(x, \varepsilon)B(x, \varepsilon)y = 0 \tag{8.11}$$

on the interval $-1 \leq x \leq 1$, where A and B are holomorphic (i.e., single valued and analytic) in x and ε and have asymptotic power series expansions

$$A(x, \varepsilon) \sim \sum_{j=0}^{\infty} A_j(x)\varepsilon^j$$

$$B(x, \varepsilon) \sim \sum_{j=0}^{\infty} B_j(x)\varepsilon^j$$

as $\varepsilon \to 0$ for x in some complex neighborhood of the interval $-1 \leq x \leq 1$ and for ε in some complex sector $S = \{0 < |\varepsilon| \leq \varepsilon_0,$

$|\arg \varepsilon| \leq \theta_0, \theta_0 > 0\}$. We shall also suppose that A and B are real and that

$$A_0(x) \neq 0 \tag{8.12}$$

for $-1 \leq x \leq 1$. The point $x = 0$, then, is a turning point of Eq. (8.11) and a singular point of the corresponding reduced equation. We shall not give a general definition of turning points. [The interested reader should refer to Wasow (1965), Olver (1973), and McHugh (1971) for general discussions of turning point theory.] We expect that our results in Section 1 where A and B were constants will be useful in obtaining the limiting behavior in the more general case.

Let us first introduce the new variable

$$\eta = \eta(x) = \left\{\frac{2}{\alpha}\int_0^x sA_0(s)\,ds\right\}^{1/2}, \tag{8.13}$$

where $\alpha = A_0(0)$ and $x\eta(x) \geq 0$. Note that $\eta'(x) = xA_0(x)/\eta(x)\alpha > 0$ for $-1 \leq x \leq 1$ and that $\eta(x)$ is a monotonically increasing function on $[-1, 1]$ with $\eta(0) = 0$. Thus x can be uniquely expressed as an increasing function of η. With respect to η, y satisfies

$$\varepsilon y_{\eta\eta} + \left[2\alpha\eta\frac{A(x,\varepsilon)}{A_0(x)} - \frac{\varepsilon\eta''(x)}{(\eta'(x))^2}\right]y_\eta - \left(\frac{\alpha^2\eta^2 A(x,\varepsilon)B(x,\varepsilon)}{x^2A_0^2(x)}\right)y = 0 \tag{8.14}$$

for $\eta_- \equiv \eta(-1) \leq \eta \leq \eta(1) \equiv \eta_+$. [We note that this equation is slightly simpler than (8.11) since the coefficient of the first derivative is now $2\alpha\eta$ when $\varepsilon = 0$.]

We shall now formally obtain

THEOREM 8: *There exist holomorphic functions $M(\eta, \varepsilon)$, $N(\eta, \varepsilon)$, and $\sigma(\varepsilon)$ such that any solution y of (8.4) can be expressed in the form*

$$y(\eta, \varepsilon) = M(\eta, \varepsilon)w + \varepsilon N(\eta, \varepsilon)w_\eta \tag{8.15}$$

where $w(\eta)$ satisfies the comparison equation

$$\varepsilon w_{\eta\eta} + 2\alpha\eta w_{\eta} - \big(\alpha\beta + \varepsilon\sigma(\varepsilon)\big)w = 0 \tag{8.16}$$

for $\beta = B_0(0)$. Here M, N, and σ have asymptotic power series expansions

$$M(\eta, \varepsilon) \sim \sum_{j=0}^{\infty} M_j(\eta)\varepsilon^j$$

$$N(\eta, \varepsilon) \sim \sum_{j=0}^{\infty} N_j(\eta)\varepsilon^j$$

$$\sigma(\varepsilon) \sim \sum_{j=0}^{\infty} \sigma_j \varepsilon^j$$

as $\varepsilon \to 0$ in S with $M(0, \varepsilon) = 1$.

Remarks

1. Note that the comparison equation (8.16) is closely related to the original equation (8.11). In particular, if $A(x, \varepsilon)$ and $B(x, \varepsilon)$ were replaced by $\alpha = A(0,0)$ and $\tilde{\beta} = \beta + \varepsilon\sigma(\varepsilon)/\alpha$ in (8.11), the equations would coincide upon substituting η for x.

As shown in Section 1, the general solution of (8.16) has the form

$$w(\eta) = \exp\left(-\frac{\alpha\eta^2}{2\varepsilon}\right)\left[C_1 D_{-1-\frac{1}{2}\tilde{\beta}}\left(\left(\frac{2\alpha}{\varepsilon}\right)^{1/2}\eta\right) + C_2 D_{\frac{1}{2}\tilde{\beta}}\left(i\left(\frac{2\alpha}{\varepsilon}\right)^{1/2}\eta\right)\right], \tag{8.17}$$

where C_1 and C_2 are arbitrary constants. Differentiating, then,

$$w_\eta(\eta) = \left(\frac{2\alpha}{\varepsilon}\right)^{1/2}\exp\left(-\frac{\alpha\eta^2}{2\varepsilon}\right)\left[-C_1 D_{-\frac{1}{2}\tilde{\beta}}\left(\left(\frac{2\alpha}{\varepsilon}\right)^{1/2}\eta\right) + \frac{i\tilde{\beta}}{2} C_2 D_{-1-\frac{1}{2}\tilde{\beta}}\left(i\left(\frac{2\alpha}{\varepsilon}\right)^{1/2}\eta\right)\right] \tag{8.18}$$

[cf. Whittaker and Watson (1952) for the appropriate differentiation formulas]. Thus the theorem implies that the general solution of (8.11) (written as a function of η) will be

$$y(\eta) = C_1 \exp\left(-\frac{\alpha\eta^2}{2\varepsilon}\right)\left[\left(M(\eta,\varepsilon)D_{-1-\frac{1}{2}\tilde{\beta}}\left(\left(\frac{2\alpha}{\varepsilon}\right)^{1/2}\eta\right)\right.\right.$$
$$\left.\left. - \varepsilon\left(\frac{2\alpha}{\varepsilon}\right)^{1/2} N(\eta,\varepsilon)D_{-\frac{1}{2}\tilde{\beta}}\left(\left(\frac{2\alpha}{\varepsilon}\right)^{1/2}\eta\right)\right)\right]$$
$$+ C_2 \exp\left(-\frac{\alpha\eta^2}{2\varepsilon}\right)\left[M(\eta,\varepsilon)D_{\frac{1}{2}\tilde{\beta}}\left(i\left(\frac{2\alpha}{\varepsilon}\right)^{1/2}\eta\right)\right.$$
$$\left. + \frac{i}{2}\varepsilon\tilde{\beta}N(\eta,\varepsilon)D_{-1-\frac{1}{2}\tilde{\beta}}\left(i\left(\frac{2\alpha}{\varepsilon}\right)^{1/2}\eta\right)\right] \quad \text{for} \quad \eta_- \le \eta \le \eta_+. \tag{8.19}$$

2. Sibuya (1962) showed that expansions analogous to those for M, N, and σ are asymptotically correct in η sectors with central angle less that π. The uniform validity of these expansions in a full neighborhood of $\eta = 0$ was established by Lee (1969). Such results are very difficult and have been achieved only for equations reducible to Airy's equation [cf. Wasow (1965, 1968)] or Weber's equation. It is likely, however, that certain new work [cf. Sibuya (1973)] will be useful in obtaining more general results.

The idea that solutions of complicated equations with turning points might be analyzed in terms of solutions of simpler equations with analogous turning points was much developed by Langer and others [see, e.g., Langer (1931, 1949), McKelvey (1955), Cherry (1950), and Wasow (1965)]. This work tends to justify the JWKB approximations widely used by physicists [cf., e.g., Olver (1973)]. For a wide variety of second-order equations, Lynn and Keller (1970) have given formal procedures for representing solutions asymptotically in terms of solutions of simpler comparison equations. Hanson and Russell (1967) have obtained related results for second-order systems.

We now give a formal proof of Theorem 8. First, rewrite (8.14) as

$$\varepsilon y_{\eta\eta} + \big(2\alpha\eta + \varepsilon\zeta(\eta,\varepsilon)\big)y_\eta - \big(\alpha\beta + 2\alpha\eta\theta(\eta) + \varepsilon\delta(\eta,\varepsilon)\big)y = 0, \tag{8.20}$$

where

$$\zeta(\eta, \varepsilon) = \frac{-\eta''(x)}{(\eta'(x))^2} + \frac{2\alpha\eta(x)}{\varepsilon A_0(x)}\Big(A(x, \varepsilon) - A_0(x)\Big),$$

$$\theta(\eta) = \frac{1}{2\eta(x)}\left(\frac{\alpha\eta^2(x) B_0(x)}{x^2 A_0(x)} - \beta\right),$$

and $\delta(\eta, \varepsilon)$ are holomorphic functions of η such that ζ and δ have power series expansions as $\varepsilon \to 0$ in S. Setting

$$y = Mw + \varepsilon N w_\eta,$$

differentiating and using Eq. (8.16), we have

$$y_\eta = \Big(M_\eta + (\alpha\beta + \varepsilon\sigma)N\Big)w + (M + \varepsilon N_\eta - 2\alpha\eta N)w_\eta.$$

Differentiating again and substituting into (8.20), we obtain

$$C(\eta, \varepsilon)w + \varepsilon D(\eta, \varepsilon)w_\eta = 0.$$

Setting C and D separately equal to zero finally implies a pair of differential equations for M and N. Thus, we have

$$\begin{aligned} C(\eta, \varepsilon) = {} & 2\alpha\eta(M_\eta - \theta M) + \varepsilon[(\sigma - \delta)M \\ & + (\alpha\beta + \varepsilon\sigma)(2N_\eta + \zeta N) + \zeta M_\eta + M_{\eta\eta}] = 0 \end{aligned} \tag{8.21a}$$

and

$$\begin{aligned} D(\eta, \varepsilon) = {} & -2\alpha\Big(\eta N_\eta + N + \eta N(\theta + \zeta)\Big) + \zeta M + 2M_\eta \\ & + \varepsilon[(\sigma - \delta)N + \zeta N_\eta + N_{\eta\eta}] = 0 \end{aligned} \tag{8.21b}$$

for $\eta_- \le \eta \le \eta_+$. When $\varepsilon = 0$, $C = 0$ implies that

$$M_{0\eta} - \theta(\eta)M_0 = 0$$

and, since $M_0(0) = 1$, we have

$$M_0(\eta) = \exp\Big[\int_0^\eta \theta(t)\,dt\Big]. \tag{8.22}$$

Further, when $\varepsilon = 0$, $D = 0$ implies that

$$\begin{aligned} 2\alpha\Big((\eta N_0)_\eta + \eta N_0(\theta(\eta) + \zeta_0(\eta))\Big) &= \zeta_0(\eta) M_0 + 2 M_{0\eta} \\ &= \Big(\zeta_0(\eta) + 2\theta(\eta)\Big) \\ &\quad \times \exp\Big[\int_0^\eta \theta(t)\,dt\Big]. \end{aligned}$$

Integrating, then,

$$2\alpha\eta N_0(\eta) = \exp\Big[\int_0^\eta \theta(t)\,dt\Big] - k \exp\Big[-\int_0^\eta \Big(\theta(t) + \zeta_0(t)\Big)\,dt\Big]$$

for some constant k. In order to avoid a pole at $\eta = 0$, however, we must select $k = 1$, so

$$N_0(\eta) = \frac{1}{2\alpha\eta}\Big[M_0(\eta) - \frac{1}{M_0(\eta)} \exp\Big[-\int_0^\eta \zeta_0(t)\,dt\Big]\Big]. \tag{8.23}$$

In general, for each $j \geq 1$, we equate coefficients of ε^j successively to zero in (8.21). Thus, M_j must satisfy

$$2\alpha\eta[M_{j\eta} - \theta(\eta) M_j] = -\sigma_{j-1} M_0 + K_{j-1}(\eta),$$

where $K_{j-1}(\eta)$ is known successively; for example,

$$K_0(\eta) = \delta_0(\eta) M_0 - \zeta_0(\eta)[M_{0\eta} + \alpha\beta N_0] - M_{0\eta\eta} - 2\alpha\beta N_{0\eta}.$$

In order to obtain a holomorphic solution $M_j(\eta)$, the right side of the differential equation must have a zero when $\eta = 0$. Thus, we must select

$$\sigma_{j-1} = K_{j-1}(0) \tag{8.24}$$

and, defining

$$L_{j-1}(\eta) \equiv \frac{1}{2\alpha\eta}[K_{j-1}(\eta) - \sigma_{j-1} M_0(\eta)],$$

we have

$$M_{j\eta} - \theta M_j = L_{j-1}.$$

Since $M_j(0) = 0$, then,

$$M_j(\eta) = \int_0^\eta L_{j-1}(t)\exp\Big[\int_t^\eta \theta(s)\,ds\Big]\,dt. \tag{8.25}$$

Further, N_j must satisfy an equation of the form

$$2\alpha\Big(\eta N_{j\eta} + N_j + \eta N_j(\theta(\eta) + \zeta_0(\eta))\Big) = \zeta_0(\eta)M_j + 2M_{j\eta} + J_{j-1}(\eta),$$

where $J_{j-1}(\eta)$ is known successively, e.g.,

$$J_0(\eta) = \Big(\sigma_0 - \delta_0(\eta)\Big)N_0 + \zeta_1(\eta)(2\alpha\eta N_0 + M_0) + \zeta_0 N_{0\eta} + N_{0\eta\eta}.$$

Thus, for N_j to be holomorphic at $\eta = 0$, we must select

$$N_j(\eta) = \frac{1}{2\alpha\eta}\int_0^\eta \Big[\Big(\zeta_0(t) + 2\theta(t)\Big)M_j(t) + 2L_{j-1}(t) + J_{j-1}(t)\Big]$$
$$\times \exp\Big[-\int_t^\eta \Big(\theta(s) + \zeta_0(s)\Big)\,ds\Big]\,dt. \tag{8.26}$$

Hence, the expansions for M, N, and σ can be determined recursively. The uniform validity of these results follows from the work of Lee (1969).

3. THE BOUNDARY VALUE PROBLEM

In this section, we wish to apply our results to the boundary value problem

$$\varepsilon y'' + 2xA(x,\varepsilon)y' - A(x,\varepsilon)B(x,\varepsilon)y = 0, \qquad -1 \le x \le 1$$
$$y(-1),\ \ y(1)\ \ \text{being prescribed constants,} \tag{8.27}$$

where A and B satisfy the same assumptions as in Section 2.

As a consequence of Theorem 8, we have:

COROLLARY 1: *Suppose* $A_0(x) > 0$ *and* $B_0(0) \equiv \beta \neq -2n$, $n = 1, 2, \ldots$. *Then the boundary value problem* (8.27) *has a unique solution* $y(x)$

for ε sufficiently small such that

$$y(x) = \begin{cases} Z_1(x) + o(1) & \text{for} \quad -1 \le x < 0 \\ O(\varepsilon^{\beta/4}) & \text{for} \quad x = 0 \\ Z_2(x) + o(1) & \text{for} \quad 0 < x \le 1, \end{cases} \tag{8.28}$$

where $Z_1(x)$ *and* $Z_2(x)$ *satisfy the reduced problems*

$$\begin{aligned} 2xZ_1' - B_0(x)Z_1 &= 0, \qquad -1 \le x < 0 \\ Z_1(-1) &= y(-1) \end{aligned} \tag{8.29}$$

and

$$\begin{aligned} 2xZ_2' - B_0(x)Z_2 &= 0, \qquad 0 < x \le 1 \\ Z_2(1) &= y(1); \end{aligned} \tag{8.30}$$

i.e.,

$$Z_{1,2}(x) = y(\mp 1)\exp\left[\int_{\mp 1}^{x} \frac{B_0(s)}{2s}\,ds\right].$$

PROOF: According to (8.19), the solution y as a function of η is of the form

$$y(\eta) = M_0(\eta)w(\eta) + o(1)$$

as $\varepsilon \to 0$ in S whenever $w_\eta = o(1/\varepsilon)$. Here

$$w(\eta) = \exp\left(-\frac{\alpha\eta^2}{2\varepsilon}\right)\left[C_1 D_{-1-\frac{1}{2}\tilde{\beta}}\left(\left(\frac{2\alpha}{\varepsilon}\right)^{1/2}\eta\right) + C_2 D_{\frac{1}{2}\tilde{\beta}}\left(i\left(\frac{2\alpha}{\varepsilon}\right)^{1/2}\eta\right)\right]$$

satisfies (8.16). Further, since $\tilde{\beta} = \beta + (\varepsilon/\alpha)\sigma(\varepsilon) \neq -2, -4, \ldots$ for ε sufficiently small, Section 1 implies

$$w(\eta) \to \begin{cases} w_1(\eta) & \text{for} \quad \eta_- \le \eta < 0 \\ O(\varepsilon^{\beta/4}) & \text{for} \quad \eta = 0 \\ w_2(\eta) & \text{for} \quad 0 < \eta \le \eta_+, \end{cases}$$

where w_1 and w_2 satisfy

$$2\eta w_{1\eta} - \beta w_1 = 0, \qquad w_1(\eta_-) = \frac{y(-1)}{M_0(\eta_-)}$$

and

$$2\eta w_{2\eta} - \beta w_2 = 0, \qquad w_2(\eta_+) = \frac{y(1)}{M_0(\eta_+)}$$

and $w_{1\eta}(\eta_-)$ and $w_{2\eta}(\eta_+)$ are bounded. Thus,

$$y(\eta) \to \begin{cases} \dfrac{y(-1)}{M_0(\eta_-)(\eta_-)^{\beta/2}} M_0(\eta)\eta^{\beta/2} & \text{for} \quad \eta_- \leq \eta < 0 \\ O(\varepsilon^{\beta/4}), & \text{for} \quad \eta = 0 \\ \dfrac{y(1)}{M_0(\eta_+)(\eta_+)^{\beta/2}} M_0(\eta)\eta^{\beta/2} & \text{for} \quad 0 < \eta \leq \eta_+ . \end{cases}$$

Note, however, that

$$\begin{aligned} \frac{M_0(\eta)}{M_0(\eta_\pm)}\left(\frac{\eta}{\eta_\pm}\right)^{\beta/2} &= \exp\left[\int_{\eta_\pm}^{\eta} \frac{1}{2\eta}\left(\frac{\alpha\eta^2 B_0}{x^2 A_0} - \beta\right) d\eta\right]\left(\frac{\eta}{\eta_\pm}\right)^{\beta/2} \\ &= \exp\left[\int_{\pm 1}^{x} \frac{\alpha\eta(x) B_0(x)}{2A_0(x)x^2}\eta'(x)\, dx\right] \\ &= \exp\left[\int_{\pm 1}^{x} \frac{B_0(s)}{2s}\, ds\right], \qquad x \neq 0 \end{aligned}$$

so the results follow.

An example illustrating Corollary 1 is given in Fig. 23. Likewise, we have

COROLLARY 2: *Suppose that $A_0(x) < 0$ and $B_0(0) \neq 2m$, $m = 0$, 1, 2, ... Then the boundary value problem (8.27) has a unique solution $y(x)$ for ε sufficiently small such that*

$$y(x) \to 0 \quad \text{as} \quad \varepsilon \to 0 \quad \text{for} \quad -1 < x < 1. \tag{8.31}$$

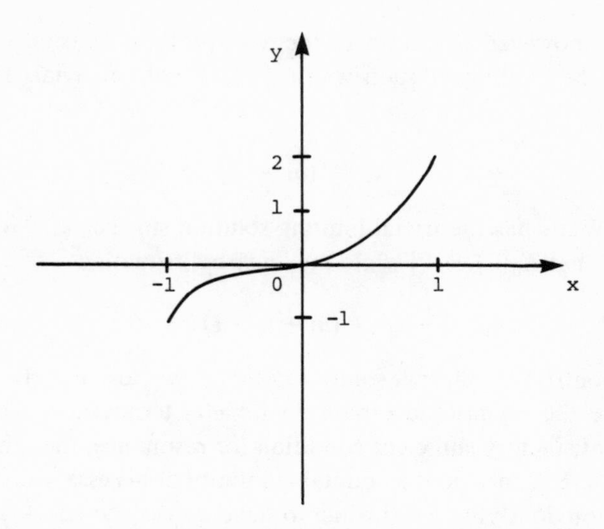

FIGURE 23 The limiting solution of $\varepsilon y'' + 2xy' + 3(1 + x^2)y = 0$, $y(-1) = -1, y(1) = 2$, ε small.

PROOF: Since all solutions of

$$\varepsilon v_{\eta\eta} + 2\alpha\eta v_\eta - \alpha(\beta + \varepsilon\sigma)v = 0$$
$$v(\eta_\pm) \quad \text{bounded}$$

and their derivatives tend to zero as $\varepsilon \to 0$ when $\alpha < 0$ and $\beta + \varepsilon\sigma \neq 0, 2, 4, \ldots$, the results follow from the theorem.

For further discussion of these results, see O'Malley (1970d).

The most interesting and obstinate case occurs when $A_0(x) < 0$ and $B_0(0) = 2m$, $m = 0, 1, 2, \ldots$. As we found when A_0 and B_0 were constants, it is then possible to obtain a nonzero limiting solution within $(-1, 1)$. Such "resonance" phenomena are potentially useful in applications. For example, boundary value problems such as (8.27) arise in the study of flow at high Reynolds number between counter-rotating disks [cf. Watts (1971)]. Ackerberg and O'Malley (1970) showed that resonance will occur if the coefficient $\sigma(\varepsilon)$ in the comparison equation (8.15) is appropriately exponentially small as

$\varepsilon \to 0$. If, however, any term in the asymptotic expansion of $\sigma(\varepsilon)$ is nonzero, the limiting solution within $(-1, 1)$ will be trivial. Thus, the example

$$\varepsilon y'' - xy' + (m + x)y = 0$$

used by Watts has the trivial limiting solution since $\sigma_0 \neq 0$ [cf. (8.24) and note that $K_0(0) \neq 0$] while the "nearby" equation

$$\varepsilon y'' - xy' + (m + x - \varepsilon)y = 0$$

allows nontrivial limiting solutions. Since we are merely able to determine the asymptotic expansion for $\sigma(\varepsilon)$ termwise, we have not given a satisfactory sufficient condition for resonance, though $\sigma_j = 0$, $j = 0, 1, 2, \ldots$, provides a countable infinity of necessary conditions. It would obviously be worthwhile to have an easy-to-check sufficient condition.

When resonance does occur, the limiting solution will satisfy the reduced equation. Thus, Ackerberg and O'Malley showed that, when $\int_{-1}^{1} xA_0(x)\,dx > 0$, the limiting solution will be

$$Z_1(x) = y(-1)(-x)^m \exp\left[\int_{-1}^{x} \left(\frac{B_0(s) - 2m}{2s}\right) ds\right]$$

throughout $-1 \leq x < 1$ while, when $\int_{-1}^{1} xA_0(x)\,dx < 0$, the limiting solution will be

$$Z_2(x) = y(1)x^m \exp\left[\int_{1}^{x} \left(\frac{B_0(s) - 2m}{2s}\right) ds\right]$$

on $-1 < x \leq 1$. When $\int_{-1}^{1} xA_0(x)\,dx = 0$, $\eta_- = -\eta_+$, and

$$y(\eta) \to M_0(\eta)v(\eta) \qquad \text{on} \quad \eta_- < \eta < \eta_+,$$

where

$$v(\eta) = \frac{1}{2}\left(\frac{v(\eta_-)}{\eta_-^m} + \frac{v(\eta_+)}{\eta_+^m}\right)\eta^m + o(1)$$

and $v_\eta(\eta_\pm) = O(1/\varepsilon)$. Thus, nonuniform convergence occurs at both endpoints and

$$y \to Cx^m \exp\left[\int_0^x \left(\frac{B_0(s) - 2m}{2s}\right) ds\right]$$

within $(-1, 1)$ for some constant $C \neq 0$ (in general). An expression for C can be easily obtained. It shows that the limiting solution will not be the average of Z_1 and Z_2 (as was incorrectly claimed by Ackerberg and O'Malley).

Attempts to solve these problems with resonance using matched asymptotic expansions have not succeeded. By combining matched expansion techniques and the WKBJ method it has been possible to obtain solutions exhibiting resonance (cf. Ackerberg and O'Malley). The results of Kreiss and Parter (1973) further indicate the subtle nature of the resonance phenomena.

A more unpleasant unbounded resonance can occur when $A_0(x) > 0$ and $B_0(0) = -2, -4, -6, \ldots$ as (8.10) suggests.

REFERENCES

ABRAHAMSSON, L. R., KELLER, H. B., and KREISS, H. O. (to appear). "Difference approximations for singular perturbations of systems of ordinary differential equations."

ACKERBERG, R. C., and O'MALLEY, R. E., JR. (1970). "Boundary layer problems exhibiting resonance," *Stud. Appl. Math.* **49**, 277–295.

ANDERSON, B. D. O. and MOORE, J. B. (1971). *Linear Optimal Control.* Prentice-Hall, Englewood Cliffs, New Jersey.

ATHANS, M., and FALB, P. L. (1966). *Optimal Control.* McGraw-Hill, New York.

BAGIROVA, N. KH., VASIL'EVA, A. B., and IMANALIEV, M. I. (1967). "The problem of asymptotic solutions of optimal control problems," *Differential Equations* **3**, 985–988.

BELLMAN, R. (1964). *Perturbation Techniques in Mathematics, Physics and Engineering.* Holt, New York.

BELLMAN, R. (1970). *Introduction to Matrix Analysis*, 2nd ed. McGraw-Hill, New York.

BJUREL, G., DAHLQUIST, G., LINDBERG, B., LINDE, S., and ODEN, L. (1970). *Survey of Stiff Ordinary Differential Equations*, Rep. Roy. Inst. of Technol., Stockholm.

BOBISUD, L. E., and CALVERT, J. (1970). "Singularly perturbed differential equations in a Hilbert space," *J. Math. Anal. Appl.* **30**, 113–127.

Boyce, W. E., and Handelman, G. H. (1961). "Vibrations of rotating beams with tip mass," *Z. Angew. Math. Phys.* **12**, 369–392.

Burghardt, A., and Zaleski, T. (1968). "Longitudinal dispersion at small and large Peclet numbers in chemical flow reactors," *Chem. Eng. Sci.* **23**, 575–591.

Carrier, G. F. (to appear). "Perturbation methods," in *Handbook of Applied Mathematics* (C. E. Pearson, ed.).

Chen, J. (1972). *Asymptotic Solutions of Certain Singularly Perturbed Boundary Value Problems Arising in Chemical Reactor Theory*. Doctoral Dissertation, New York Univ., New York.

Chen, J., and O'Malley, R. E., Jr. (1974). "On the asymptotic solution of a two-parameter boundary value problem of chemical reactor theory," *SIAM J. Appl. Math.* **26**.

Cherry, T. M. (1950). "Uniform asymptotic formulas for functions with transition points," *Trans. Amer. Math. Soc.* **68**, 224–257.

Cochran, J. A. (1962). *Problems in Singular Perturbation Theory*. Doctoral Dissertation, Stanford Univ., Stanford, California.

Cochran, J. A. (1968). "On the uniqueness of solutions of linear differential equations," *J. Math. Anal. Appl.* **22**, 418–426.

Coddington, E. A., and Levinson, N. (1952). "A boundary value problem for a nonlinear differential equation with a small parameter," *Proc. Amer. Math. Soc.* **3**, 73–81.

Coddington, E. A., and Levinson, N. (1955). *Theory of Ordinary Differential Equations*. McGraw-Hill, New York.

Cohen, D. S. (1972). "Multiple solutions of singular perturbation problems," *SIAM J. Math. Anal.* **3**, 72–82.

Cohen, D. S. (1973a). "Multiple solutions of nonlinear partial differential equations," *Lecture Notes in Mathematics* **322**, 15–77. Springer-Verlag, Berlin and New York.

Cohen, D. S. (1973b), "Singular perturbations of nonlinear two-point boundary value problems," *J. Math. Anal. Appl.* **43**, 151–160.

Cole, J. D. (1951). "On a quasi-linear parabolic equation occuring in aerodynamics," *Quart. Appl. Math.* **9**, 225–236.

Cole, J. D. (1968). *Perturbation Methods in Applied Mathematics*. Ginn (Blaisdell), Boston, Massachusetts.

Collins, W. D. (to appear). "Singular perturbations of linear time-optimal control problems."

Cooke, K. L. and Meyer, K. R (1966). "The condition of regular degeneration of singularly perturbed systems of linear differential–difference equations," *J. Math. Anal. Appl.* **14**, 83–106.

COURANT, R., and HILBERT, D. (1962). *Methods of Mathematical Physics*, Vol. II, *Partial Differential Equations*. Wiley (Interscience), New York.

DI PRIMA, R. C. (1968). "Asymptotic methods for an infinitely long slider squeeze-film bearing," *J. Lubric. Technol.* **90**, 173–183.

DESOER, C. A., and SHENSA, M. J. (1970). "Networks with very small and very large parasitics: Natural frequencies and stability," *Proc. IEEE* **58**, 1933–1938.

DICKEY, R. W. (1973). "Inflation of the toroidal membrane," *Quart. Appl. Math.* **31**, 11–26.

DORR, F. W., PARTER, S. V., and SHAMPINE, L. F. (1973). "Application of the maximum principle to singular perturbation problems," *SIAM Rev.* **15**, 43–88.

ECKHAUS, W. (1973). *Matched Asymptotic Expansions and Singular Perturbations*. North Holland Publ., Amsterdam.

ERDÉLYI, A. (1956). *Asymptotic Expansions*. Dover, New York.

ERDÉLYI, A. (1961a). "An expansion procedure for singular perturbations," *Atti. Accad. Sci. Torino, Cl. Sci. Fis. Mat. Natur.* **95**, 651–672.

ERDÉLYI, A. (1961b). "General asymptotic expansions of Laplace integrals," *Arch. Ration. Mech. Anal.* **7**, 1–20.

ERDÉLYI, A. (1964). "The integral equations of asymptotic theory," in *Asymptotic Solutions of Differential Equations and Their Applications* (C. Wilcox, ed.) pp. 211–229. Wiley, New York.

ERDÉLYI, A. (1968a). "Approximate solutions of a nonlinear boundary value problem," *Arch. Ration. Mech. Anal.* **29**, 1–17.

ERDÉLYI, A. (1968b). "Two-variable expansions for singular perturbations," *J. Inst. Math. Appl.* **4**, 113–119.

ERDÉLYI, A. and WYMAN, M. (1963). "The asymptotic evaluation of certain integrals," *Arch. Ration. Mech. Anal.* **14**, 217–258.

FIFE, P. C. (1973). "Semilinear elliptic boundary value problems with small parameters," *Arch. Ration. Mech. Anal.*

FIFE, P. C. (1974). "Transition layers in singular perturbation problems," *J. Differential Equations* **15**.

FRAENKEL, L. E. (1969). "On the method of matched asymptotic expansions," *Proc. Cambridge Phil. Soc.* **65**, 209–284.

FRIEDMAN A., (1969). "Singular perturbations for the Cauchy problem and for boundary value problems," *J. Differential Equations* **5**, 226–261.

FRIEDMAN, B. (1969). *Lectures on Applications-Oriented Mathematics*. Holden-Day, San Francisco, California.

FRIEDRICHS, K. O. (1950). "Kirchhoff's boundary conditions and the edge effect for elastic plates," *Proc. Symp. Appl. Math.*, 1950, **3**, 117–124. McGraw-Hill, New York.

FRIEDRICHS, K. O. (1953). *Special Topics in Analysis*, lecture notes. New York Univ., New York.

FRIEDRICHS, K.O. (1955). "Asymptotic phenomena in mathematical physics," *Bull. Amer. Math. Soc.* **61**, 367–381.

FRIEDRICHS, K. O., and STOKER, J. J. (1941). "The nonlinear boundary value problem of the buckled plate," *Amer. J. Math.* **63**, 839–888.

GILBARG, D. (1951). "The existence and limit behavior of the one-dimensional shock layer," *Amer. J. Math.* **73**, 256–274.

GOL'DENVEIZER, A. L. (1961). *Theory of Elastic Thin Shells*. Pergamon Press, New York.

GORDON, N. (1973). *Matched Asymptotic Expansion Solutions of Cauchy Problems with Small Parameters*. Doctoral Dissertation, New York Univ., New York.

HABER, S., and LEVINSON, N. (1955). "A boundary value problem for a singularly perturbed differential equation," *Proc. Amer. Math. Soc.* **6**, 866–872.

HADDAD, A. H., and KOKOTOVIĆ, P. V. (1971). "A note on singular perturbation of linear state regulators," *IEEE Trans. Automatic Control* **16**, 279–281.

HADLOCK, C. R. (1970). *Singular Perturbations of a Class of Two Point Boundary Value Problems Arising in Optimal Control*. Doctoral Dissertation, Univ. of Illinois, Urbana.

HADLOCK, C., JAMSHIDI, M. and KOKOTOVIĆ, P. (1970). "Near optimal design of three-time scale systems," *Proc. Annu. Conf. Inform. Sc. and Syst., 4th*, Princeton, 1970, 118–122.

HALE, J. K. (1969). *Ordinary Differential Equations*. Wiley (Interscience), New York.

HANDELMAN, G. H., KELLER, J. B., and O'MALLEY, R. E., JR. (1968). "Loss of boundary conditions in the asymptotic solution of linear ordinary differential equations," *Commun. Pure Appl. Math.* **21**, 243–261.

HANSON, R. J., and RUSSELL, D. L. (1967). "Classification and reduction of second order systems at a turning point," *J. Math. Phys. (Cambridge, Mass.)* **46**, 74–92.

HARRIS, W. A., JR. (1960). "Singular perturbations of two-point boundary problems for systems of ordinary differential equations," *Arch. Ration. Mech. Anal.* **5**, 212–225.

HARRIS, W. A., JR. (1961). "Singular perturbations of eigenvalue problems," *Arch. Ration. Mech. Anal.* **7**, 224–241.

HARRIS, W. A., JR. (1962). "Singular perturbations of a boundary value problem for a system of differential equations," *Duke Math. J.* **29**, 429–445.

HARRIS, W. A., JR. (1965). "Equivalent classes of singular perturbation problems," *Rend. Circ. Mat. Palermo* **14**, 61–75.

HARRIS, W. A., JR. (1973). "Singularly perturbed boundary value problems revisited." *Lecture Notes in Mathematics* **312**, 54–64. Springer-Verlag, Berlin and New York.

HEINEKIN, F. G., TSUCHIYA, H. M., and ARIS, R. (1967). "On the mathematical status of the pseudo-steady state hypothesis of biochemical kinetics," *Math. Biosci.* **1**, 95–173.

HOPPENSTEADT, F. C. (1966). "Singular perturbations on the infinite interval," *Trans. Amer. Math. Soc.* **123**, 521–535.

HOPPENSTEADT, F. (1970). "Cauchy problems involving a small parameter," *Bull. Amer. Math. Soc.* **76**, 142–146.

HOPPENSTEADT, F. (1971). "Properties of solutions of ordinary differential equations with a small parameter," *Commun. Pure Appl. Math.* **24**, 807–840.

HOPPENSTEADT, F. (1974). "A two-locus model with linkage and recombination: Slow selection case."

HUKUHARA, M. (1937). "Sur les propriétés asymptotiques des solutions d'un système d'équations différentielles linéares contenant un paramètre," *Mem. Fac. Eng. Kyushu Imp. Univ.* **8**, 249–280.

HUTCHINSON, G. E. (1948). "Circular causal systems in ecology," *Ann. N. Y. Acad. Sci.* **50**, 221–246.

JAMESON, A., and O'MALLEY, R. E., JR. (1974). "Singular perturbations and singular arcs."

KALMAN, R. E., FALB, P. L., and ARBIB, M. A. (1969). *Topics in Mathematical Systems Theory*. McGraw-Hill, New York.

KAPLUN, S. (1967). In *Fluid Mechanics and Singular Perturbations* (P. A. Lagerstrom, L. N. Howard, and C. S. Liu, eds.). Academic Press, New York.

KASYMOV, K. A. (1968). "The initial jump problem for nonlinear systems of differential equations containing a small parameter," *Sov. Math. Dokl.* **9**, 345–349.

KATO, T. (1966). *Perturbation Theory for Linear Operators*. Springer-Verlag, Berlin and New York.

KELLER, H. B. (1973). "Stability theory for multiple equilibrium states of a nonlinear diffusion process: A singularly perturbed eigenvalue problem," *SIAM J. Math. Anal.* **4**, 134–140.

KELLER, J. B., and LEWIS, R. M. (to be published). *Asymptotic Theory of Wave Propagation and Diffraction.*

KEVORKIAN, J. (1966). "The two-variable expansion procedure for the approximate solution of certain nonlinear differential equations," in *Lectures in Applied Mathematics*, 7, *Space Mathematics*, (J. B. Rosser, ed.) Part III, pp. 206–275. Amer. Math. Soc., Providence, Rhode Island.

KOKOTOVIĆ, P. V. (1972). "A control engineer's introduction to singular perturbations," *Singular Perturbation: Order Reduction in Control System Design*, pp. 1–12. Amer. Soc. Mech. Eng., New York.

KOKOTOVIĆ, P. V., and SANNUTI, P. (1968). "Singular perturbation method for reducing the model order in optimal control design," *IEEE Trans. Automatic Control* **13**, 377–384.

KOKOTOVIĆ, P. V., and YACKEL, R. A. (1972). "Singular perturbation theory of linear state regulators: Basic theorems," *IEEE Trans. Automatic Control* **17**, 29–37.

KOLLETT, F. W. (1973). "Two-timing methods on expanding intervals," *SIAM J. Math. Anal.*

KREIN, S. G. (1971). *Linear Differential Equations in Banach Space* (transl. of 1963 Russ. ed.). Amer. Math. Soc., Providence, Rhode Island.

KREISS, H. O., and PARTER, S. V. (1973). "Remarks on singular perturbations with turning points," *SIAM J. Math. Anal.*

KUNG, C-F (1973). *On a Singular Perturbation Problem in Optimal Control Theory*, Doctoral Dissertation, New York Univ., New York.

LAGERSTROM, P. A., and CASTEN, R. G. (1972). "Basic concepts underlying singular perturbation techniques," *SIAM Rev.* **14**, 63–120.

LANGER, R. E. (1931). "On the asymptotic solution of ordinary differential equations with an application to the Bessel functions of larger order," *Trans. Amer. Math. Soc.* **33**, 23–64.

LANGER, R. E. (1949). "The asymptotic solutions of ordinary linear differential equations of the second order with special reference to a turning point," *Trans. Amer. Math. Soc.* **67**, 461–490.

LATTA, G. E. (1951). *Singular Perturbation Problems.* Doctoral Dissertation, California Inst. Tech., Pasadena.

LEE, R. Y. (1969). "On uniform simplication of linear differential equations in a full neighborhood of a turning point," *J. Math. Anal. Appl.* **27**, 501–510.

LEVIN, J. J., and LEVINSON, N. (1954). "Singular perturbations of nonlinear systems of differential equations and an associated boundary layer equation," *J. Ration. Mech. Anal.* **3**, 247–270.

Levinson, N. (1951). "Perturbations of discontinuous solutions of nonlinear systems of differential equations," *Acta Math.* **82**, 71–106.

Lions, J. L. (1972). *Some Aspects of the Optimal Control of Distributed Parameter Systems*, CBMS Regional Conf. Ser. in Appl. Math., SIAM Publ., Philadelphia, Pennsylvania.

Lions, J. L. (1973). "Perturbation singulierès dans les problèmes aux limites et en contrôle optimal," *Lecture Notes in Mathematics* **323**. Springer-Verlag, Berlin and New York.

Lynn, R. Y. S., and Keller, J. B. (1970). "Uniform asymptotic solutions of second order linear ordinary differential equations with turning points," *Comm. Pure Appl. Math.* **23**, 379–408.

McHugh, J. A. M. (1971). "An historical survey of ordinary differential equations with a large parameter and turning points," *Arch. History Exact Sci.* **7**, 277–324.

McKelvey, R. W. (1955). "The solution of second order linear ordinary differential equations about a turning point of order two," *Trans. Amer. Math. Soc.* **79**, 103–123.

Macki, J. W. (1967). "Singular perturbations of a boundary value problem for a system of nonlinear ordinary differential equations," *Arch. Ration. Mech. Anal.* **24**, 219–232.

Mahony, J. J. (1961–1962). "An expansion method for singular perturbation problems," *J. Australian Math. Soc.* **2**, 440–463.

Meerov, M. V. (1961). *Introduction to the Dynamics of Automatic Regulating of Electrical Machines* (transl. of 1956 Russ. ed.). Butterworth, London.

Meyer, R. E. (1967). "On the approximation of double limits by single limits and the Kaplun extension theorem," *J. Inst. Math. Appl.* **3**, 245–249.

Meyer, R. E. (1971). *Introduction to Mathematical Fluid Dynamics*. Wiley, New York.

Miranker, W. L. (1973). "Numerical methods of boundary layer type for stiff systems of differential equations," *Computing*.

Morrison, J. A. (1966). "Comparison of the modified method of averaging and the two-variable expansion procedure," *SIAM Rev.* **8**, 66–85.

Morse, P. M., and Feshbach, H. (1953). *Methods of Theoretical Physics*, Vol. II. McGraw-Hill, New York.

Murphy W. D. (1967). "Numerical analysis of boundary layer problems in ordinary differential equations," *Math. Comp.* **21**, 583–596.

Murray, J. D. (1973). "On Burgers' model equations for turbulence," *J. Fluid*

Mech. **59**(2), 263–279.

NAU, R. W., and SIMMONDS, J. G. (1973). "Calculation of low natural frequencies of clamped cylindrical shells by asymptotic methods," *Internat. J. Solids Structures* **9**, 591–605.

NAYFEH, A. H. (1973). *Perturbation Methods*. Wiley, New York.

OLVER, F. W. J. (1973). *Asymptotics and Special Functions*. Academic Press, New York.

O'MALLEY, R. E., JR. (1967a) "Two-parameter singular perturbation problems for second order equations," *J. Math. Mech.* **16**, 291–308.

O'MALLEY, R. E., JR. (1967b). "Singular perturbations of boundary value problems for linear ordinary differential equations involving two parameters," *J. Math. Anal. Appl.* **19**, 291–308.

O'MALLEY, R. E., JR. (1968a). "A boundary value problem for certain nonlinear second order differential equations with a small parameter," *Arch. Ration. Mech. Anal.* **29**, 66–74.

O'MALLEY, R. E., JR. (1968b) "Topics in singular perturbations," *Adv. Math.* **2**, 365–470. Reprinted in *Lectures on Ordinary Differential Equations* (R. W. McKelvey, ed.), pp. 155–260. Academic Press, New York, 1970.

O'MALLEY, R. E., JR. (1969a). "On a boundary value problem for a nonlinear differential equation with a small parameter," *SIAM J. Appl. Math.* **17**, 569–581.

O'MALLEY, R. E., JR. (1969b). "Boundary value problems for linear systems of ordinary differential equations involving many small parameters," *J. Math. Mech.* **18**, 835–856.

O'MALLEY, R. E., JR. (1969c). "On the asymptotic solution of boundary value problems for nonhomogeneous ordinary differential equations containing a parameter," *J. Math. Anal. Appl.* **28**, 450–460.

O'MALLEY, R. E., JR., (1970a). "A nonlinear singular perturbation problem arising in the study of chemical flow reactors," *J. Inst. Math. Appl.* **6**, 12–21.

O'MALLEY, R. E., JR. (1970b). "On singular perturbation problems with interior nonuniformities," *J. Math. Mech.* **19**, 1103–1112.

O'MALLEY, R. E., JR. (1970c). "Singular perturbation of a boundary value problem for a system of nonlinear differential equations," *J. Differential Equations*, **8**, 431–447.

O'MALLEY, R. E., JR. (1970d). "On boundary value problems for a singularly perturbed equation with a turning point," *SIAM J. Math. Anal.* **1**, 479–490.

O'MALLEY, R. E., JR. (1971a). "On initial value problems for nonlinear systems of differential equations with two small parameters," *Arch. Ration. Mech. Anal.* **40**, 209–222.

O'Malley, R. E., Jr. (1971b). "On the asymptotic solution of initial value problems for differential equations with small delay," *SIAM J. Math. Anal.* **2**, 259–268.

O'Malley, R. E., Jr. (1972a). "The singularly perturbed linear state regulator problem," *SIAM J. Contr.* **10**, 399–413.

O'Malley, R. E., Jr. (1972b). "Singular perturbation of the time-invariant linear state regulator problem," *J. Differential Equations* **12**, 117–128.

O'Malley, R. E., Jr. (1972c). "On multiple solutions of a singular perturbation problem," *Arch. Ration. Mech. Anal.* **49**, 89–98.

O'Malley, R. E., Jr. (1972d). "On the asymptotic solution of certain nonlinear singularly perturbed optimal control problems," *Singular Perturbations: Order Reduction in Control System Design*, pp. 44–56. Amer. Soc. of Mech. Eng., New York.

O'Malley, R. E., Jr. (1974). "On two methods of solution for a singularly perturbed linear state regulator problem," *SIAM Review*.

O'Malley, R. E., Jr. and Keller, J. B. (1968). "Loss of boundary conditions in the asymptotic solution of linear differential equations. II. Boundary value problems," *Commun. Pure Appl. Math.* **21**, 263–270.

O'Malley, R. E., Jr. and Mazaika, P. K. (1971). "On the asymptotic solution of multi-point boundary value problems with discontinuous coefficients," *Indiana Univ. Math. J.* **20**, 667–682.

Parter, S. V. (1972). "Remarks on the existence theory for multiple solutions of a singular perturbation problem," *SIAM J. Math. Anal.* **3**, 496–505.

Perko, L. M. (1969). "Higher order averaging and related methods for perturbed periodic and quasi-periodic systems," *SIAM J. Appl. Math.* **17**, 698–724.

Pittnauer, F. (1969). "Holomorphic functions with prescribed asymptotic expansions," *SIAM J. Appl. Math.* **17**, 607–613.

Pittnauer, F. (1972) "Vorlesungen über asymptotische Reihen," *Lecture Notes in Mathematics* **301**, Springer-Verlag, Berlin and New York.

Prandtl, L. (1905). "Über Flüssigkeits-bewegung bei kleiner Reibung," *Verh. Int. Math. Kongr. 3rd*, pp. 484–491. Tuebner, Leipzig.

Protter, M. H., and Weinberger, H. F. (1967). *Maximum Principles in Differential Equations*. Prentice-Hall, Englewood Cliffs, New Jersey.

Lord Rayleigh (J. W. Strutt), (1945). *The Theory of Sound*, 2nd ed., Vol. 1, Dover, New York (original printing 1894).

Raymond, L. R., and Amundson, N. R. (1964). "Some observations on tubular reactor stability," *Can. J. Chem. Eng.* **42**, 173–177.

Reiss, E. L. (1962). "On the theory of cylindrical shells," *Quart. J. Mech. Appl.*

Math. **15**, 325–338.

Reiss, E. L. (1971). "On multivariable asymptotic expansions," *SIAM Rev.* **13**, 189–196.

Rellich, F. (1969). *Perturbation Theory of Eigenvalue Problems.* Gordon & Breach, New York.

Riekstins, E. Y. (1966). "On the use of neutrices for asymptotic representation of some integrals," *Latvian Math. Annu.* pp. 5–21.

Sannuti, P. (1971). "Asymptotic expansions of singularly perturbed quasilinear optimal systems," *Proc. Annu. Conf. Circuit and Syst. Theory, 9th, Allerton,* 1971, pp. 182–191. Univ. of Illinois, Urbana, Illinois.

Sannuti, P., and Kokotović, P. V. (1969). "Near optimum design of linear systems by a singular perturbation method," *IEEE Trans. Automat. Contr.* **14**, 15–22.

Sannuti, P., and Reddy, P. B. (1973). "Asymptotic series solution of optimal systems with small time-delay," *IEEE Trans. Automat. Contr.* **18**, 250–259.

Schmidt, H. (1937). "Beiträge zu einer Theorie des allgemeinen asymtotischen Darstellungen," *Math. Ann.* **113**, 629–656.

Searl, J. W. (1971). "Expansions for singular perturbations," *J. Inst. Math. Appl.* **8**, 131–138.

Sibuya, Y. (1958). "Sur réduction analytique d'un système d'equations différentielles ordinaires linéaires contenant un paramètre," *J. Fac. Sci. Univ. Tokyo Sect. 1*, **7**, 527–540.

Sibuya, Y. (1962). "Asymptotic solutions of a system of linear ordinary differential equations containing a parameter," *Funkcial. Ekvac.* **4**, 83–113.

Sibuya, Y. (1973). "*Uniform Simplification in a Full Neighborhood of a Turning Point.*" Report #1320, Mathematics Research Center, Univ. of Wisconsin, Madison.

Stengle, G. (1971). "A new basis for uniform asymptotic solution of differential equations containing one or several parameters," *Trans. Amer. Math. Soc.* **156**, 1–43.

Trenogin, V. A. (1963). "Asymptotic behavior and existence of a solution of the Cauchy problem for a first order differential equation with a small parameter in a Banach space," *Sov. Math. Dokl.* **4**, 1261–1265.

Trenogin, V. A. (1970). "The development and application of the asymptotic method of Lyusternik and Vishik," *Russ. Math. Surveys* **25**, 119–156.

Turrittin, H. L. (1936). "Asymptotic solutions of certain ordinary differential equations associated with multiple roots of the characteristic equation," *Amer. J. Math.* **58**, 364–378.

Turrittin, H. L. (1952). "Asymptotic expansions of solutions of systems of ordinary linear differential equations containing a parameter," *Contrib. Theory Nonlinear Oscillations* **2**, 81–116.

Vainberg, M., and Trenogin, V. A. (1962). "The methods of Lyapunov and Schmidt in the theory of nonlinear equations and their further development," *Russ. Math. Surveys* **17**, 1–60.

van der Corput, J. G. (1955–1956). "Asymptotic developments I. Fundamental theorems of asymptotics," *J. Anal. Math.* **4**, 341–418.

Van Dyke, M. (1964). *Perturbation Methods in Fluid Dynamics*. Academic Press, New York.

Van Dyke, M. (1971). "Nineteenth-century applications of the method of matched asymptotic expansions," *Conference on Singular Perturbation Techniques, Princeton Univ., January 1971*. Unpublished.

Vasil'eva, A. B. (1962). "Asymptotic solutions of differential-difference equations in the case of a small deviation in the argument," *USSR Comput. Math. Math. Phys.* **2**, 869–893.

Vasil'eva, A. B. (1963a). "Asymptotic behavior of solutions to certain problems involving nonlinear differential equations containing a small parameter multiplying the highest derivatives," *Russ. Math. Surveys* **18**, 13–84.

Vasil'eva, A. B. (1963b). "Asymptotic methods in the theory of ordinary differential equations containing small parameters in front of the higher derivatives," *USSR Comput. Math. Math. Phys.* **3**, 823–863.

Vasil'eva, A. B. (1969). *Asymptotic Expansions of Solutions of Singularly Perturbed Problems*, Rep. 1077. Math. Res. Center, Univ. of Wisconsin, Madison.

Vasil'eva, A. B. (1972). "Asymptotic solution of two-point boundary value problems for singularly perturbed conditionally stable systems," *Singular Perturbations: Order Reduction in Control System Design*, pp. 57–62. Amer. Soc. of Mech. Eng., New York.

Vasil'eva, A. B., and Tupciev, V. A. (1960). "Asymptotic formulae for the solution of a boundary value problem in the case of a second order equation containing a small parameter in the term containing the highest derivative," *Sov. Math. Dokl.* **1**, 1333–1335.

Vishik, M. I., and Lyusternik, L. A. (1957). "Regular degeneration and boundary layer for linear differential equations with a small parameter," *Usp. Mat. Nauk.* **12**, 3–122 [*Amer. Math. Soc. Transl.* [2], **20**, 239–364 (1961)].

Vishik, M. I., and Lyusternik, L. A. (1958). "On the asymptotic behavior of the solutions of boundary value problems for quasi-linear differential equations," *Dokl. Akad. Nauk. SSSR* **121**, 778–781.

VISHIK, M. I., and LYUSTERNIK, L. A. (1960). "Initial jump for nonlinear differential equations containing a small parameter," *Sov. Math. Dokl.* **1**, 749–752.

WALKER, R. J. (1962). *Algebraic Curves*, reprint. Dover, New York.

WASOW, W. (1941). *On Boundary Layer Problems in the Theory of Ordinary Differential Equations.* Doctoral Dissertation, New York Univ. New York.

WASOW, W. (1944). "On the asymptotic solution of boundary value problems for ordinary differential equations containing a parameter," *J. Math. Phys. (Cambridge, Mass.)* **23**, 173–183.

WASOW, W. (1956). "Singular perturbation of boundary value problems for nonlinear differential equations of the second order," *Commun. Pure Appl. Math.* **9**, 93–113.

WASOW, W. (1964). "Asymptotic decomposition of systems of linear differential equations having poles with respect to a parameter," *Rend. Circ. Mat. Palermo* **13**, 329–344.

WASOW, W. (1965). *Asymptotic Expansions for Ordinary Differential Equations.* Wiley (Interscience), New York.

WASOW, W. (1968). "Connection problems for asymptotic series," *Bull. Amer. Math. Soc.* **74**, 831–853.

WASOW, W. (1970). "The capriciousness of singular perturbations," *Nieuw Arch. Wisk.* **18**, 190–210.

WATTS, A. M. (1971). "A singular perturbation problem with a turning point," *Bull. Australian Math. Soc.* **5**, 61–73.

WHITHAM, G. B. (1970). "Two-timing, variational principles, and waves," *J. Fluid Mech.* **44**, 373–395.

WHITTAKER, E. T. and WATSON, G. N. (1952). *A Course of Modern Analysis*, 4th ed. Cambridge Univ. Press, London and New York.

WILLETT, D. (1964). "Nonlinear Volterra integral equations as contraction mappings," *Arch. Ration. Mech. Anal.* **15**, 79–86.

WILLETT, D. (1966). "On a nonlinear boundary value problem with a small parameter multiplying the highest derivative," *Arch. Ration. Mech. Anal.* **23**, 276–287.

YACKEL, R. A., and WILDE, R. R. (1972). "Boundary layer methods for the optimal control of singularly perturbed linear systems," *Singular Perturbations: Order Reduction in Control System Design*, pp. 13–36. Amer. Soc. of Mech. Eng., New York.

YARMISH, J. (1972). *Aspects of the Numerical and Theoretical Treatment of Singular Perturbations*, Doctoral Dissertation, New York Univ., New York.

INDEX